Science in History

J. D. Bernal

In 4 volumes

Science in History

Volume 1: The Emergence of Science

100 illustrations

The M.I.T. Press
Cambridge, Massachusetts

First published by C. A. Watts & Co. Ltd, 1954
Third edition 1965
Illustrated edition published simultaneously
by C. A. Watts & Co. Ltd and in Pelican Books 1969
Copyright © J. D. Bernal, 1965, 1969

Designed by Gerald Cinamon

First M.I.T. Press Paperback Edition, March 1971
Second printing, April 1974
Third printing, January 1977
Fourth printing, April 1979

ISBN 0 262 02073 4 (hardcover)
ISBN 0 262 52020 6 (paperback)

Library of Congress catalog card number: 78–136489

Printed and bound in the United States of America

Contents

Contents

Preface

In 1948 I was asked to deliver the Charles Beard Lectures at Ruskin College, Oxford. The subject I chose was 'Science in Social History'. It was one that had interested me for many years and there seemed no difficulty in presenting it to an intelligent and unspecialized audience. When I came to give the lectures, and still more when I undertook to present them in book form, I began to realize that I had opened a subject that required far more study and hard thinking than I had given it up till then. It was, however, one far too fascinating to put down, and I decided to persevere in it. The first result of that intention is this book, one which I hoped to prepare in three weeks but which has already taken me twice that number of years. It is only now that I am beginning to understand what the problems are of the place of science in history.

Scientists in the past were able to neglect all but their immediate predecessors' work and even to reject the traditions of the past as more likely to block than assist progress. Now, however, the troubles of the times, together with the inescapable connexion between them and the advance of science, have focused attention on the historical aspect of science. To find how to overcome the difficulties that face us and to release the new forces of science for welfare rather than destruction, it is necessary to examine anew how the present situation came about.

In the last thirty years, largely owing to the impact of Marxist thought, the idea has grown that not only the means used by natural scientists in their researches but also the very guiding ideas of their theoretical approach are conditioned by the events and pressures of society. This idea has been violently opposed and as energetically supported; but in the controversy the earlier view of the direct impact of science on society has become overshadowed. It was my purpose to emphasize once more to what extent the advance of natural science has helped to determine that of society itself; not only in the economic changes brought about by the application of scientific discoveries, but also by the effect on the general frame of thought of the impact of new scientific theories.

I soon found, however, that this involved far more than drawing up a catalogue of inventions and hypotheses and illustrating by examples how these affected economic and political developments. This had been done often enough already. If anything new and significant was to be hoped for nothing less would be adequate than a complete re-examination of the reciprocal relations of science and society. It would be as one-sided to assess the effects of science on society as of society on science.

Nor would it be enough to confine the inquiry to recent times. This might have sufficed if all that was sought was the effects of the material changes in the pattern of life brought about in the Industrial Revolution and at an accelerated pace ever since. But if in addition it was necessary to seek to discover how the advance of science had altered the whole frame of human thought, it would also be necessary to go back through the great controversies of the Renaissance about the nature of the heavens, and then still farther back to the Ancients, without whose theories the controversies would have no meaning.

There was nothing for it but to attempt to trace the whole story from the very origins of human society. This involved a parallel study of all social and economic history in relation to the history of science, a task well beyond the scope of any individual, even of those who have devoted their whole lives to historical studies. For a busy scientist untrained in the techniques of historical research it would be sheer presumption to attempt a serious full-scale analysis and presentation of this aspect of history. Yet there seemed some excuse for making a first attempt to sketch out the field, if only to stimulate, through its omissions and errors, others more leisured and better qualified to produce a more authoritative picture. Moreover, there was a compensating advantage in the position of a working scientist who has lived long enough to have followed through, and even participated in, the scientific movements of critical periods, both of science and of social change. I have indeed been exceptionally fortunate in having first-hand experience in the carrying out and organization of scientific work, and in seeing it called for and used for practical purposes both in peace and in war.

It is in the light of that experience that I have attempted to evaluate the conditions and attitudes that have prevailed inside and outside science in other times. No attempt is made here to present a chronologically uniform picture. The present century has witnessed such an immense upsurge of science and has seen that science used so rapidly and to such effect – to cite no other examples than penicillin and the atom bomb – that consideration of the development of twentieth-century science has required a good half of the book. Here the scientist of the day is in as good a position as the historian, and every reader can criticize from his own experience.

Science throughout is taken in a very broad sense and nowhere do I attempt to cramp it into a definition. Indeed, science has so changed its nature over the whole range of human history that no definition could be made to fit. Although I have aimed at including everything called science, the centre of interest of the book lies in natural science and technology because, for reasons that will be discussed, the sciences of society were at first embodied in tradition and ritual and only took shape under the influence and on the model of the natural sciences. The theme which constantly recurs is the complex interaction between techniques, science, and philosophy. Science stands as a middle term between the established and transmitted practice of men who work for their living, and the pattern of ideas and traditions which assure the continuity of society and the rights and privileges of the classes that make it up.

Science, in one aspect, is ordered technique; in another, it is rationalized mythology. Because it started as a hardly distinguishable aspect of the mystery of the craftsman and the lore of the priest, which remained separate over most of recorded history, science was long in establishing any independent existence in society. Even when it did find its own specific adepts in medicine, astrology, and alchemy, these formed for many ages a small group parasitic on wealthy princes, clerics, and merchants. It is only in the last three centuries that science has become traditionally established as a profession in its own right, with its specific education, literature, and fellowship. Now, in our own time, we are witnessing a beginning of a return to the earlier state of humanity through a general pervasion of science into all forms of practical activity and thought, bringing together once more the scientist, the worker, and the administrator.

The progress of science has been anything but uniform in time and place. Periods of rapid advance have alternated with longer periods of stagnation and even of decay. In the course of time the centres of scientific activity have been continually displaced, usually following rather than leading the migration of the centres of commercial and industrial activity. Babylonia, Egypt, and India have all been the foci of ancient science. Greece became their common heir, and there the rational basis for science as we know it was first worked out. That forward movement of human thought came to an end even before the final decay of the classical city States. There was little place for science in Rome and none in the barbarian kingdoms of western Europe. The heritage of Greece returned to the East from which it had come. In Syria, Persia, and India, even in far-away China, new breaths of science stirred and came together in a brilliant synthesis under the banner of Islam. It was from this source that science and techniques entered medieval Europe. There they underwent a development which, though slow at first, was to give rise to the great outburst of creative activity which resulted in modern science.

An unbroken and active tradition links us with the revolutionary science of the Renaissance, though we can distinguish in its development four major periods of advance. The first, centred in Italy, produced the renewal of mechanics, anatomy, and astronomy with Leonardo, Vesalius, and Copernicus, destroying the authority of the Ancients in their central doctrines of man and the world. The second, spreading now to the Low Countries, France, and Britain, beginning with Bacon, Galileo, and Descartes, and ending in Newton, hammered out a new mathematical–mechanical model of the world. After an interval, the third transformation, centred in industrial Britain and revolutionary Paris, opened to science areas of experience, such as that of electricity, untouched by the Greeks. It was then that science could help in a decisive way with power, machinery, and chemicals, to transform production and transport. The fourth and greatest of all in extent and effect, if not in intrinsic intellectual performance, is the scientific revolution of our own time. We are witnessing the beginning of a world science, transforming old and creating new industries, permeating every aspect of human life. It is now also, during this period of transition, that we find science directly involved in the violent and terrible drama of wars and social revolution.

It is by now apparent that each of these great periods of science corresponds to one of social and economic change. Greek science reflects the rise and decline of the money-dominated, slave-owning iron age society. The long interval of the Middle Ages marked the growth and instability of feudal subsistence economy with little use for science. It was not until the bonds of feudal order were broken by the rise of the bourgeoisie that science could advance. Capitalism and modern science were born in the same movement. The phases of the evolution of modern science mark the successive crises of capitalist economy. The first two periods coincide with its early battles and its first success in establishing itself as the dominant economy in Holland and Britain. The third ushered in the factory system and seemed to forecast the triumph of a progressive capitalism allied to science. By the time the last had come capitalism was already overgrown and overreaching itself, and the new form of socialism was visibly struggling to replace it and to take over, in order to use in its own way the now proved forces of science.

To write this, however, is only to begin to state the problem. These rough equations between social and scientific development give rise to one central question. How in detail does a social transformation affect science? What gave the science of ancient Athens, of Renaissance Florence, of eighteenth-century Birmingham and Glasgow, its particular drive and novelty? How, conversely, did the achievements of the scientists of those times and places affect the industry, the commerce, the politics, and religion of their contemporaries? How much of that effect was permanent and how much a passing fashion? These are questions which I have examined and attempted to answer.

I have tried, in doing this, to take into account as many as I could of the relevant factors. I have tried to determine and describe the technical possibilities and limitations of each period, the degree of economic incentive for urging on and fixing securely the advances that were made. But advances are not made by impersonal forces, but by living men and women. Their lives and livelihoods, their motives, the relations with the political movements of the time, had all to be considered. It was necessary to estimate from their works and writings how far they were stimulated, or how far retarded, by the ideas they drew from old traditions or from the active controversies of their times.

At every turn this conflict between the forces tending to advance and those tending to retard science comes into prominence. We can perceive the positive progressive forces breaking through at the beginning of each critical advance, and the regressive forces of pedantry and obscurantism reasserting themselves at its close. Yet in each case the circumstances are different and require a separate examination.

It would be absurd to expect to find any simple explanations of the critical phases of the development of science. Nevertheless merely to bring out the connexions between social, technical, and scientific factors should be enough to lead to further study and to a deeper, even if unformulated understanding. I know myself that this returning to the past has inevitably coloured my comprehension of the present and my ideas of the future path of science. True, in science, more perhaps than in any other field of human enterprise, progress is possible, and has indeed largely occurred, without any knowledge of history; but that knowledge is bound to affect the future direction and course of science and, if the lessons of the past have been well read, progress will be quicker and surer.

This book represents a first attempt to put down in order some of these lessons of the past. It is not, nor is it intended to be, another history of science, though it must needs set out again much of that history and refer to more. Its aim is to bring out the influence of science upon other aspects of history, whether direct or indirect, through its effect on economic changes, or through its influence on the ideas of the ruling classes of the day or of those who are striving to supplant them. But, as will be seen, these influences are rarely clear-cut, nor are they usually one-way influences. Often enough the ideas which the statesmen and divines think they have taken from the latest phase of scientific thought are just the ideas of their class and time reflected in the minds of scientists subjected to the same social influences. Certainly much of the influence of Newton and Darwin in Britain was of that character, but this did not prevent them from being revolutionary when they were presented elsewhere against a different social background.

The more I followed up the social historical interactions of science the closer knit they appeared. I began to see something of the size and intricacy of the task I had attempted and the absolute impossibility of presenting at the same time a fully convincing and intelligible picture. If I did not put in enough I should be accused of imposing ready-made solutions; if I put in too much the reader would lose the clue in a mass of detail. I have sought the best compromise I could find, but what I have managed to produce is admittedly less well documented and less closely argued than the finished work I originally planned. It will succeed in the measure that the reader can follow the course of the history traced out. Rather than assenting to any particular conclusion of mine, I would wish him to look at history in a new way, to make his own discoveries and frame his own theories.

Length and time imposed severe limitations. I must write a book, not an encyclopedia, and I must bring it to an end in a finite number of years. These, and the fact that I have never been able to find any continuous stretch of time for writing but have had to take it up and drop it at odd intervals, are responsible for some of its defects, of which no one can be more aware than I. I know that the history is full of omissions and errors in detail that could have been put right had I had the time and the scholarship to discover and deal with them. I hope vigilant readers will point them out and not dismiss the whole work because in some field in which they have a special competence they have found me straying. What I must hope is that these errors as to established facts, as well as other errors which stem from gaps in the record, will not radically affect the validity of the theses I am sustaining. No scientist can be, nor can he seriously want to be, guaranteed against reversals of judgement in the long run. All he can hope for, as I do, is to establish enough valid and significant connexions between facts, even if they are later overthrown, to serve as a basis for finding new facts and new connexions.

The plan of the book was originally determined by that of the lectures from which it grew, but each lecture became first a chapter and has then swollen into a part containing a number of chapters. The introductory chapter (Part 1, Chapter 1) is one in which the major problems are stated, and there is some discussion in general on the nature and method of science and on its place in society. Because of its somewhat abstract character non-scientists might be advised to leave it until after reading the historical and descriptive portions. Those contained in Parts 2, 3, 4, and 5, making up the first half of these books, deal with the whole range of history from the dawn of human society to the eve of the twentieth century. Part 2, Chapters 2, 3, and 4, deals with the emergence of science from its forerunners in technique and social custom to its full formulation in the hands of the Greeks. In Part 3, Chapters 5 and 6 deal with the recovery and slow growth of a science and technology, through Islam and Christendom, to the end of the Middle Ages.

Part 4, which contains only Chapter 7, deals with the birth of modern science in the great revolutionary epoch of the Renaissance. It ends in the seventeenth century with a renewed science closely linked with a young and assertive capitalism. Part 5, Chapters 8 and 9, is mainly the record of the spread of an established science and its share in the transformation of industry in an era of capitalist domination up to the illusory golden age of the end of the nineteenth century.

Part 6 is nearly all devoted to the twentieth century and largely to contemporary science and politics. It is divided not by time, but by subject. Chapter 10 deals with the physical sciences, with the growth of the electrical and chemical industries, and with the culminating achievement, for ill or good, of the hydrogen bomb. Chapter 11 deals with the biological sciences and their impact on agriculture, medicine, and warfare. Chapters 12 and 13 enter the disputed field of the social sciences, which for continuity needs to be traced back beyond the confines of the century. In all the historic chapters, 2–13, the plan is first to present a picture of social and scientific development of each successive period, and then to bring out the relations between them. Part 7, Chapter 14, attempts to summarize and draw conclusions, with an eye to the future, from the whole of history.

The scope is evidently comprehensive, but to obtain the results aimed at this is necessary. A partial account would miss the point of presenting the total picture, for it would inevitably fail to question what is taken for granted in what is omitted. Even to leave out remote and uncertain origins would not do, for, as I hope to show, much of what is obscure and difficult in the science of our times and in its social context depends on attitudes and institutions passed down from those times.

No more need be written here. The book itself is the only test of whether I have succeeded in what I set out to do and to what degree it was worth doing.

J.D.B.
London, April 1954

Preface to the Third Edition

The seven years that have passed since the printing of the second edition of *Science in History* have made necessary changes far greater than were needed for the second edition. The extent of these changes are themselves a sign of the times, in which both the range of human knowledge and the rate of its application to social life have given ample evidence that we are passing into a new era. In fact it is highly unlikely that another edition of this book on the basis of the present one will ever be produced. The changes that have occurred in recent years are such that a complete recasting of our appreciation of the importance and growth of science will be required.

It became evident already, in making the second edition, that to treat the twentieth century as a whole was becoming illogical. It was then beginning to appear that we had entered by 1940, with the war, a new era characterized at the outset by the discovery of nuclear fission and its use in the atomic bomb. In the twenty years since then, the Atomic Age has been enlarged to the Space Age. With it have come the less spectacular, but probably more important, practical applications of electronics to computation and automation. These changes have affected the science, the economics and the politics not only of a corner of Europe and America, as in the past, but of the whole world.

I have not attempted to deal specifically with the new era in the present edition, partly because of the enormous difficulties it would entail and partly because its character is still not sufficiently marked to give a coherent account of it by itself. We have seen so many fundamental changes in all fields of knowledge and practice in these years, that it will be inevitable that their analysis will bring about a complete revision of man's attitude towards society and the world of nature. But, though new discoveries have been made and new applications carried out, this profound adaptation of human thought has not yet taken place. It would be prudent to give time for further analysis before attempting a separate description of this period.

Instead of doing this I have taken the whole of Part 6, 'Science in our Time', and have almost completely rewritten it. Changes have been made in the other chapters owing to the rapid advance of the history of science itself, but these are comparatively minor. I have included also, in the final chapter, 14, a special section on the effects of the enormous growth of science in the last decade and the increasing rapidity of its advance.

The essential character of the new changes is not to be measured merely by their external manifestations. The scientific-technical revolution, which could be fairly seen as far back as the thirties, is now recognized outside the world of science, particularly in politics, as the dominant feature of our time. Science is now what it certainly was not in earlier centuries, a necessity to the mere survival of the human race and at the same time the largest step in advance the human race has ever had to make. During the last decade man has achieved the fantastic adventure of leaving the planet and wandering in space: the whole scope of man's understanding of the outer universe has been enormously enlarged.

Great as the conquest of space may be, the advances in the last few years in the study of the minutiae of matter and life are far greater, both in themselves as intellectual achievements and in their effect on human life. The first part of the twentieth century had witnessed the birth and the development of our knowledge of the physical atoms, culminating in the mid-century in the double-edged discovery of the utilization of atomic power and of nuclear bombs. This search is by no means ended: we have still to find a proper way of using thermonuclear power for peace and not for war.

With this has gone such an improvement, indeed transformation, of industrial processes, chemical even more than physical, that it has in principle relieved man from dependence on the wasting natural resources of the earth. Material abundance is here for the taking. The other great by-product of physical science has been the development of faster and larger computers and the discovery of the logic of their use in all human occupations involving skill and judgement, in administration as well as production. We have seen only the beginning of this revolution in automation, and already, in the advanced industrial countries such as the United States, it has produced a progressive and apparently irreversible pattern of unemployment, first affecting the manual workers in mass-production factories and now reaching into the office and even into the Board Room. Even if these ill effects are temporary – and I think they will be – the computer and the automatic factory have come to stay. They represent an enormous eventual liberation of the human mind as well as human body from heavy and dull tasks.

Unfortunately, this extraordinarily rapid evolution of understanding in techniques moves much faster than the social arrangements that control them. The new methods of production inevitably require social adaptations to cope with them successfully, and this fact is being discovered at the moment, especially in the old industrial countries. Socialist countries were already aware of it and their remarkable progress is largely due to their appreciation of the growing importance of science in its social context.

Within a few years mankind, first of all in the industrialized countries and then throughout the world, will be as much affected by these scientific and technical changes as by any changes that have occurred in previous history. The technical and scientific revolution of our days is of an importance transcending the earlier revolutions which brought agriculture and mechanical production: it implies great changes in the whole pattern of human life – a much greater emphasis on education and on training in how to use and to enjoy the new powers. This in turn will have a profound social effect. The old concept, which goes at least as far back as ancient Egypt, of an educated *élite* and a mass of illiterate peasants and workers, is bound to disappear, indeed is disappearing already. The archaic methods of control of production and consumption sanctified in the code of free-enterprise capitalism, which has become in reality monopoly capitalism, will have to make way for planned production and to make more and more use of mathematical and computational methods. In simple words, science implies socialism.

Great as have been these changes they are likely to be dwarfed by those that are coming now. In the second half of the twentieth century the great revolution has been in biology; not in one branch or another, but in the common fusing together of all branches of biology, from genetics to molecular structure. Discoveries have enabled the bridge to be made between what can be seen and felt in biological processes, and the mutual positions of the very atoms that comprise them. Discoveries are beginning to show how the nucleic acids, the carriers of heredity, transfer the information built into them, according to a code locked in the chromosomes, to the formation of the specific enzyme proteins which carry out the current living processes. The discovery and elaboration of these mechanisms, which has only just begun, should completely transform our understanding of life, making precise what was previously vague or incomprehensible, and by that means open up these processes to voluntary control. This is already being exercised in innumerable medical channels as well as in agriculture. The concept of the evolution of whole organisms introduced by Darwin has now been pushed back to the consideration of the evolution of the common molecular processes which occur in all life. The organic unity of all terrestrial life may now be considered as proved. This raises a new major problem as to how does this single chemical pattern maintain itself in its multiple forms, ranging from bacteria to human beings and oak trees.

This points again to the general convergence of all sciences in this century. It touches the problem of the origin of life itself and its relation to the origin of planets, stars and galactic systems. The connexion between microcosm and macrocosm has been transferred from the mystic imaginations of astrology to experimental, controllable facts.

Similar but not so dramatic discoveries have characterized the other boundary of biology, that which concerns itself with control and communications – the evolution of brain, of habit and consciousness of animals. This leads to consideration of the special social evolution of humanity itself, the noosphere of that imaginative thinker Teilhard de Chardin. For the first time in human history we can hope to trace precisely the whole field of knowledge from nebulae to politics. This general picture has already revealed a characteristic structure, one first found in outer space by the astronomer Charlier but really applying all the way through nature. We find everywhere a system of box within box of units, aggregating at a certain stage to form larger units which can then aggregate in turn. For example, gas and dust form stars, stars form clusters, clusters form galaxies, galaxies form galactic clusters and meta-galaxies. In an analogous way, organisms are composed of organs composed of tissues, composed of cells with organelles built from characteristic macromolecules such as the nucleic acids. All these are arrangements that exist not only in space but also in time. Each complex appears at a specific stage in its own evolution, but not everywhere at the same rate, for new stars are being formed today and organisms existed two or three thousand million years ago.

Our new knowledge is not, however, in any sense absolute, quite the contrary. We know better now what we do not know. But that is not an expression of scepticism, it is a programme for action. The provisional nature of science was never more evident. We have to learn at the same time how to carry on intelligent action knowing that we do not know. This demands the creation of what is effectively a new comprehensive branch of science, a real *science of science*, combining psychological, historical and material factors that lead to discovery and that will be needed in the planning of science.

The planning of science is, indeed, one of the most important characteristics of this latest phase of the scientific-industrial revolution. Long opposed by the old school of academic scientists it is now in full swing and is bound to spread on both the national and the international planes.

The success of planned science has tended to obscure the contrast between our ability to use the new knowledge in a practical way and our inability to comprehend it deeply. This is because it is felt in capitalist countries that to do so is certain to affect and even threatens to destroy long-established and revered philosophical and religious ideas. Yet this result can no more be avoided than were the effects of the earlier revolutions in science.

Science itself will be profoundly affected. The effects of the convergence of scientific disciplines and their overlapping into the social and economic fields are tending to create a unified science. This is not on the basis of any positivistic reinterpretations of words but on that of a deep thinking out of the implications of discoveries and the means of attaining them. This in turn is bound to affect the application of science, which will gradually spread over the whole of human activities.

These great changes in science and industry have moreover taken place in a period of social and economic transformation affecting the whole world. With growing speed and in the midst of struggles and atrocities unmatched in the past, the domination of imperialism, at least in its colonial form, is being successfully challenged. The use of the techniques inherited from the first industrial revolution had enabled the industrial nations of Europe and later of North America to take effective political and economic control of the whole world. This process of spreading imperialism is now being reversed and already the end of political colonialism is in sight. With its completion the whole population of the world will come to achieve the high standard of living of the formerly privileged fifth. This will mean multiplying by five the numbers of people affected by and ultimately contributing to the new scientific revolution.

The peoples of Asia and Africa and Central and South America are now entering, not as select *élites* but as masses, into the effective world of our time. But they have to enter a world torn by political divisions and armed with annihilating weapons. They are entering it precisely at a time when, owing to the first effects of the scientific revolution, the control of death has been achieved while the control of births has not. The world population is growing at such a rate that it is often referred to as 'the population explosion'. The benefits of science are very poorly spread. The food provided by subsistence agriculture is only a small fraction of that which could be produced on the same land and with far fewer people by the use of modern techniques and chemical fertilizers.

Actually the difference between the standards of living of the peoples in the developed and in the under-developed world is not yet diminishing: it is increasing in a way that seems bound to lead to a crisis, and there is always the danger that a crisis of this sort might itself set off a first nuclear world war. This crisis can and must be avoided, but it can be avoided only by the efforts of the people themselves of all countries. For that they must get the necessary education and find the capital to build themselves a scientific and industrial complex which can provide for their needs.

In doing this we shall have, for the first time, the possibility of having one world at a level of intercommunicating culture and production mechanisms and one in which *all* are able to contribute to and enjoy the advance of science.

I have dealt with these subjects at greater length in this edition at the expense of omitting a certain amount of historical treatment of the social sciences. In general, I may sum up the change between this and earlier editions as one of extension of field, both scientifically and humanly, combined with a far greater degree of integration of different fields of science and of science with the processes of industry and politics.

J.D.B.
1965

Preface to the Illustrated Edition

In 1965, the third edition of *Science in History*, dealing with the inter-relations between science and society, was published. It was bound to need changing with the rapid development of both. Its main lines now require bringing up to date more than ever before. This is the object of this preface to the illustrated edition.

The scientific-technical revolution is in full swing and this is now generally recognized, but so far only in words. Its practical effect has yet to be fully appreciated and used. Already, however, it has produced effects which have disturbed world economics and politics. The tendencies discussed in Parts 6, 7 and 8 are more than ever apparent. In particular, the great gap between the developed world and the underdeveloped world, far from closing, is widening rapidly. While science is playing a larger and larger part in the advanced industrial countries, it is stagnant or even receding in those parts of the world which contain the bulk of its population. The effect of this is to bring about for the first time the possibility that humanity will extinguish itself by war or famine. Science, as it is now being used, contributes to making such a horrifying prospect not only possible but almost certain, and up till now there has been little evidence of factors which will cause this process to reverse. This vast prospect of Nemesis, however imminent, has caused little alarm and produced virtually no efforts to deal with it. It would seem there is a universal tacit conspiracy to avoid thinking about it by those responsible for creating the situation in the advanced countries, and the victims' complaints are met with indifference and repression.

The great adventure of science seems, very sadly, to lead to such an end as negates all its original promise through the ages. Even now, although the resources spent on science are greater than they have ever been, most of them are being used for preparing and carrying out destruction. There is no space here to examine these facts but they are sufficiently clearly adumbrated in the book itself, the general optimistic tone of which is borne out by the past successes of science. What has made things go astray has been the association of science with the very same forces that caused its growth, namely, its links with the factors of capitalism and imperialism.

In this book, the history of science in the nineteenth and twentieth centuries is examined to bring out its *potential*, to tell us what can be done with the new powers as they are developed, how they can be used, and to show that they are not actually being so used. It would appear that the destructive aspect of applied science is determined by the possibilities of profit and its attendant manifestations of prestige and aggression.

It is the coincidence of these anticipated events that make the task of reprinting *Science in History* both timely and difficult. I have shown that the progress of science is both physically and biologically closely tied up with its use for war in nuclear weapons and biological poisons. The shape of science has been heavily distorted by this; but another process is also taking place. The scientific advances ordered – or at least paid for – by military interests are rapidly changing the face of war and making it even more cruel than it has ever been before. The effect is not limited to physical destruction. The mental effects of lying propaganda help to complete the process and react on the modern scientific aggressors themselves, producing a brutalization and distortion of scientific values on a scale beyond past imagination of horrors demonstrated in the calm discussions of 'megadeaths' and 'overkill'. The men of the century of Big Science are being reduced to the worst kind of savage imaginable. Moreover, behind them lies a population whose values have been mentally corrupted by the policy and ideology of anti-communism.

These are general considerations. To examine the book in more detail and to bring out particularly those aspects that are new since the third edition, the following paragraphs may be useful. The first two volumes covering the time up to the Renaissance do not require much alteration. New discoveries in archaeology and history do not alter the general picture, but they do push the dates of the development of civilization further and further back. The great discovery of the prosperous town of 8000 B.C. at Lipesky Var on the Danube has shaken the theory of the Asiatic origin of agriculture, and the presence of portrait studies on stone must revise the whole history of art. There are probably some more surprises in store and the continuity of techniques and ideas that were later to give rise to science has been reaffirmed by new work on the Renaissance itself. We have to revise and give greater credit to the work of Leonardo da Vinci, whose last notebooks have been found after 300 years, wrongly catalogued, in Madrid. Other discoveries have shown that Newton's work on the solar system was taken directly from Galileo.

The Industrial Revolution and its relation to the development of science is arousing much new interest. We are interested in the first industrial revolution not only in itself, but as the basis on which the second scientific-industrial revolution of our time has been built. These aspects have been discussed in a number of recent books among which the most important are Galbraith's *New Industrial State* and Servan-Schreiber's *Le Défi américain*. Our most modern industries are frankly based on recent developments in science, mainly in physics, electronics and control mechanisms. They are leading to automation and to its far greater application to management and administration. The firms concerned are now largely in the hands of a few United States corporations which are busy extending their control over all similar firms throughout the world and thus gaining control of all the key points of modern industry. Evolved as most of them were as auxiliaries to the war industries, particularly the aerospace industry, they constituted a concerted trio to control all the older forms and thus to master, physically, chemically and biologically, the whole world. It is difficult to see how this process can be arrested. The old capitalist powers, such as England, France and Germany, cannot by themselves, on account of their small size and outdated methods, hope to do so. Nor are they capable of effective union. Further, no single country possesses a large enough market to ensure a really autonomous modern industrial complex and, therefore, no single country can compete with American industry and Big Science; nor is it provided with adequate profit motives to check the American invasion.

It is now evident that the real source of wealth lies no longer in raw materials, the labour force or machines, but in having a scientific, educated, technological manpower base. Education has become the real wealth of the new age. Unfortunately, governments are reluctant to spend enough money on education and students are not attracted to scientific or technological studies, seeing more immediate advance and prestige in the traditional arts studies. Thus the American infiltration proceeds almost unchecked, at least from outside the United States. It remains to be seen how far the checks imposed by the President on U.S. investment abroad are effective in halting the process.

These are the problems facing the western capitalist states, but, all taken together, these comprise less than one-third of the world population. Leaving aside the Socialist world, we have the so-called Third World, who are not troubled with solving the problems of affluence but rather with the eternal problems of poverty, now much aggravated by the population explosion.

If the industrial states themselves cannot manage to build up an autonomous industry, what chance have the countries of the Third World to do so? The 'aid' which is being offered to them is plainly inadequate in quantity and is hedged with so many conditions that it takes more out of the countries than it gives them. Ideally, such countries, particularly those emerging from colonialism, should rely exclusively on their own resources of materials and men. In other words, they will have to pull themselves up by their own boot-straps. Nevertheless, to some extent they may receive genuine aid from the Socialist countries. Whether in fact they can succeed in the short time before famine conditions are upon them will depend upon the efforts of the peoples of the countries themselves and their genuine friends throughout the world.

The importance of science for history has now become a matter for government concern. It is being discussed widely in many countries and is the subject for international conferences. One aspect was discussed at the U.N. Conference on the Application of Science and Technology for the Benefit of the Less Developed Areas (Geneva, 1963) and, if it did not result in substantial help from science for the underdeveloped countries, showed at least that the subject was of interest and concern. In Britain, we have initiated the Science of Science Foundation which is actively discussing the kind of problems I have been considering in this preface and, indeed, has published a book on the subject entitled *The Science of Science*.

The original purpose of *Science in History* has now been fulfilled but its task has not ended. The period of the century which has witnessed the discoveries of atomic power, space navigation and electronic computers and has given us at its end the code of genetic reproduction, marks clearly not an end but a beginning. It is for the peoples of the world to insist that this new knowledge is used in the interests of human well-being.

J.D.B.
February 1968

Acknowledgements

This book would have been impossible to write without the help of many of my friends and of my colleagues on the staff of Birkbeck College, who have advised me and directed my attention to sources of information.

In particular 1 would like to thank Dr E. H. S. Burhop, Mr Emile Burns, Professor V. G. Childe, Mr Maurice Cornforth, Mr Cedric Dover, Mr R. Palme Dutt, Dr W. Ehrenberg, Professor B. Farrington, Mr J. L. Fyfe, Mr Christopher Hill, Dr S. Lilley, Mr J. R. Morris, Dr J. Needham, Dr D. R. Newth, Dr M. Ruhemann, Professor G.Thomson and Dona Torr. They have seen and commented on various chapters of the book in its earlier stages, and I have attempted to rewrite them in line with their criticisms. None, however, have seen the final form of the work and they are in no sense responsible for the statements and views I express in it.

I would like also especially to thank my secretary, Miss A. Rimel, and her assistants, Mrs J. Fergusson and Miss R. Clayton, for their help in the technical preparation of the book – a considerable task, as it was almost completely rewritten some six times – and its index.

My thanks are also due to the librarians and their staffs at the Royal Society, The Royal College of Physicians, The University of London, Birkbeck College, The School of Oriental and African Studies, and the Director and Staff of the Science Museum, London.

Finally, I would like to record my gratitude to my assistant, Mr Francis Aprahamian, who has been indefatigable in searching for and collecting the books, quotations and other material for the work and in correcting manuscripts and proofs. Without his help I could never have attempted a book on this scale.

J.D.B.
1954

Acknowledgements to the Illustrated Edition

For the preparation of this special illustrated edition of *Science in History*, I must thank, first of all, Colin Ronan, who chose the illustrations and wrote the captions.

I should also like to thank Anne Murray, who has been responsible for correlating all the modifications involved in producing a four-volume version and for correcting the proofs.

Finally, I thank my personal assistant, Francis Aprahamian, who advised the publishers at all stages of the production of this edition.

J.D.B.
1968

Note

In the first edition of this book, I avoided the use of footnotes. A few notes have been added to subsequent editions and are marked with an asterisk (*) or a dagger (†) (if there is more than one footnote on a page). The notes have been collected together at the end of each volume and are referred to by their page numbers.

The reference numbers in the text relate to the bibliography, which is also to be found at the end of each volume. The bibliography has eight parts that correspond to the eight parts of the book. Volume 1 contains Parts 1–3; Volume 2 contains Parts 4 and 5; Volume 3 contains Part 6; Volume 4 contains Parts 7 and 8.

Part 1 of the bibliography is divided into three sections. The first contains books that cover the whole work, including general histories of science. The second section contains histories of particular sciences and the books relevant to Part 1. The third section lists periodicals to which reference has been made throughout the book.

Parts 2, 3, 4 and 5 of the bibliography are each divided into two sections. The first section in each case contains the more important books relevant to the part, and the second the remainder of the books.

In Part 6 of the bibliography, the first section contains books covering the introduction and Chapter 10, the physical sciences; and the second section, Chapter 11, the biological sciences.

Part 7 of the bibliography contains books covering the introduction and Chapters 12 and 13, the social sciences.

Part 8 of the bibliography contains books covering Chapter 14, the conclusions.

The system of reference is as follows: the first number refers to the part of the bibliography; the second to the number of the book in that part; and the third, when given, to the page in the book referred to. Thus 2.5.56 refers to page 56 of the item numbered 5 in the bibliography for Part 2, i.e. Farrington's *Science in Antiquity*.

The Emergence
and Character of Science

Chapter 1

Introduction

This book is an attempt to describe and to interpret the relations between the development of science and that of other aspects of human history. Its ultimate object is to lead to an understanding of some of the major problems which arise from the impact of science on society. Civilization as we know it today would, in its material aspects, be impossible without science. In its intellectual and moral apects science has been as deeply concerned. The spread of scientific ideas has been a decisive factor in re-moulding the whole pattern of human thought. Especially do we find in the conflicts and aspirations of our time a continual and growing involvement of science. Men live in fear of destruction by the atom bomb or biological weapons; in hope of living better lives through the application of science in agriculture and medicine. The two camps into which the world is now divided exemplify different objectives in the use of science. The urgency to reconcile them is also in part due to the catas-trophic and suicidal nature of scientific warfare.

The march of events brings before us, ever more insistently, problems about science such as: the proper use of science in society, the militariza-tion of science, the relations of science to governments, scientific secrecy, the freedom of science, the place of science in education and general cul-ture. How are such problems to be solved? Attempts to solve them by appeals to accepted principles or self-evident truths have led so far only to confusion. They can give no clear answer, for instance, as to the res-ponsibility of the scientist to the tradition of science, to humanity, or to the State. In a rapidly changing world little can be expected from ideas taken unaltered from a society that has vanished beyond recall. But this is not to say that the problems are insoluble, and in consequence to lapse into the impotent pessimism and irrationality that are so characteristic of intellectuals in capitalist countries today.[1.60] Ultimately these prob-lems must be solved and will be answered in practice in the process of finding the way of using and developing science most harmoniously and with the best results for humanity. Already much experience has been gained in countries where science has been consciously devoted to the tasks of construction and welfare. Even in Britain and America the experience of the use of science in war and war preparation has taught scientists something of what could be done in peace.[1.2.295]

But experience by itself is not enough, and indeed it can never operate alone. Consciously or unconsciously it is bound to be guided by theories and attitudes drawn from the general fund of human culture. In so far as it is unconscious, this dependence on tradition will be blind and will lead only to the repetition of attempted solutions that changed conditions have made unworkable. In so far as it is conscious, it must involve a deeper knowledge of the whole relation of science to society, for which the first requirement is the knowledge of the history of science and of society. In science, more than in any other human institution, it is necessary to search out the past in order to understand the present and to control the future.

Such an assertion would, at least until recently, have received scant support from working scientists. In natural science, and especially in the physical sciences, the idea is firmly held that current knowledge takes the place of and supersedes all the knowledge of the past. It is admitted that future knowledge will in turn make present knowledge obsolete, but for the moment it is the best available knowledge. All useful earlier knowledge is absorbed in that of the present; what has been left out are only the mistakes of ignorance. Briefly, in the words of Henry Ford, 'History is bunk.'

Fortunately more and more scientists in our time are beginning to see the consequences of this attitude of neglect of history, and with it, necessarily, of any intelligent appreciation of the place of science in society. It is only this knowledge that can prevent the scientists, for all the prestige they enjoy, being blind and helpless pawns in the great contemporary drama of the use and misuse of science. It is true that in the recent past scientists and people at large got on very nicely in the comfortable belief that the application of science led automatically to a steady improvement in human welfare. The idea is not a very old one. It was a revolutionary and dangerous speculation in the days of Roger Bacon (p. 306) and was first confidently asserted by Francis Bacon (p. 441) 300 years later. It was only the immense and progressive changes in science and manufacture that came about with the Industrial Revolution that were to make this idea of progress an assured and lasting truth – almost a platitude (p. 1085)– in Victorian times. It is certainly not so now, in these grim and anxious days, when the power that science can give is seen to be more immediately capable of wiping out civilization and even life itself from the planet than of assuring an uninterrupted progress in the arts of peace. Though even here doubt has crept in and some neo-Malthusians fear that even curing disease is dangerous on an overcrowded planet.

Whether for good or ill the importance of science today needs no emphasizing, but it does, just because of that importance, need understanding. Science is the means by which the whole of our civilization is rapidly being transformed. And science is growing; not, as in the past, steadily and imperceptibly, but rapidly, by leaps and bounds, for all to see. The fabric of our civilization has already changed enormously in our own lifetimes and is changing more and more rapidly from year to year. To understand how this is taking place it is not sufficient to know what science is doing now. It is also essential to be aware of how it came to be what it is, how it has responded in the past to the successive forms of society, and how in its turn it has served to mould them.

Some people take it for granted that, because science is affecting our lives more and more, it follows that the scientists themselves are in effective control of the mechanism of civilization and that in consequence they are immediately and largely responsible for the evils and disasters of our time. Most of those actually working in science know well enough how far this belief is from the truth. The use to which the work of the scientists is put is almost completely out of their hands. The responsibility of the scientists remains, therefore, a purely moral one. Even that responsibility is usually evaded in the tradition of science by the exaltation of the disinterested search for truth, irrespective of any consequences that may arise from it. This convenient evasion of responsibility, as we shall see (pp. 663 f.), worked well enough as long as general social progress, largely thanks to science, seemed to be the order of the day. Then the scientist could identify himself reasonably easily with the current economic and political trends and be happy enough to be left alone to pursue his freely chosen path. But in the face of a world of increasing want, misery, and fear, and one too where science itself is more and more directly involved in the more unpleasant aspects of warfare, this attitude is beginning to break down. The moral responsibility of the scientist in the world of today is difficult to evade.

The alternative is not irresponsibility, but a more conscious and active social responsibility where, on the one hand, science can make an explicit contribution to the planning of industry, agriculture, and medicine, for ends of which the scientist can fully approve; and where, on the other hand, science can be so extended and transformed as to become an integral part of the life and work of all.

The change from a socially irresponsible to a socially responsible science is one which is only just beginning. Its nature and directives are not yet fully formulated. It is only one aspect, though a vital one, of the major social transformations from an economy motivated by individual acquisitiveness to one directed to common welfare. This is going to be one of the most momentous changes in the whole of human history, and hence it is of the utmost importance that it should be fully debated and well understood in advance, for it contains great dangers as well as unlimited possibilities. It is the need to achieve this transformation in the best way, and to secure the intelligent utilization of science at every stage in it, that is the strongest reason for the study of the relations of science and society in the past, for only through this study can it be adequately understood.

ASPECTS OF SCIENCE

Before beginning this inquiry something must be said on the meaning and scope of science itself. Now of course it might seem most natural and convenient to start with a definition of science. Professor Dingle, in his extensive review [1.53] of my book *The Social Function of Science*, demands that this should be done. According to him, the writer should begin

by identifying this phenomenon and delineating as clearly as possible what it was in itself, apart from any function it might have or any relation in which it might stand to other phenomena; and he would then proceed to consider the part it played, or could play, in social life.

My experience and knowledge have convinced me of the futility and emptiness of such a course. Science is so old, it has undergone so many changes in its history, it is so linked at every point with other social activities, that any attempted definition, and there have been many, can only express more or less inadequately one of the aspects, often a minor one, that it has had at some period of its growth. Einstein [1.55] has put this point in his own way:

Science as something existing and complete is the most objective thing known to man. But science in the making, science as an end to be pursued, is as subjective and psychologically conditioned as any other branch of human endeavour — so much so, that the question 'what is the purpose and meaning of science?' receives quite different answers at different times and from different sorts of people.

To a human activity which is itself only an inseparable aspect of the unique and unrepeatable process of social evolution, the idea of definition does not strictly apply[1.5] (p. 1232).

More than any other human occupation, science is, by its very nature, changeable. It is also, as one of the latest achievements of humanity, changing most rapidly. Nor has it long had a separate existence. At the dawn of civilization it was only one aspect of the work of the magician, the cook, or the smith. It was not until the seventeenth century that it began to achieve an independent status; and that independence may itself be only a temporary phase. In the future it may well be that scientific knowledge and method will so pervade all social life that science will once again have no distinct existence. Since a definition is intrinsically impossible, the only way of conveying what is being discussed in this book as science will need to be an extensive and unfolding description. This will be the task of the later chapters, but here, as a clue to the more detailed treatment, is an attempt to analyse in a few words the major aspects in which science appears in the contemporary world.

Science may be taken, (1.1) as an institution; (1.2) as a method; (1.3) as a cumulative tradition of knowledge; (1.4) as a major factor in the maintenance and development of production; and (1.5) as one of the most powerful influences moulding beliefs and attitudes to the universe and man. In 1.6 the interactions of science and society are discussed. By listing these different aspects of science I do not intend to imply that there are as many different 'sciences'. With any concept so wide-ranging in time, connexion and category, multiplicity of aspect and reference must be the rule. The word science or scientific has a number of different meanings according to the context in which it is used. Professor Dingle took the trouble to list ten of these taken from my book. In one case cited by him science is being contrasted with engineering, a matter of the degree of practical application; in another, the scientific method as a means of verification is contrasted with the intuitive recognition of discovery. All are significant uses of the word science, but to extract the full meaning from them, they need to be linked together in a general picture of the development of science. Of the aspects listed above, those of science as an institution and as a factor in production belong almost exclusively to modern times. The method of science and its influence on beliefs date back to Greek times, if not earlier. The tradition of knowledge passed on from parent to child, from master to apprentice, is the very root of science, existing from the earliest ages of man and long before science could be considered as an institution, or could have evolved a method distinct from common sense and traditional lore.

1.1 Science as an Institution

Science as an institution in which hundreds of thousands of men and women find their profession is a very recent development. It is only in the twentieth century that the profession of science has come to compare in importance with the far older professions of the Church and the Law. It is also being recognized as something distinct from, though allied to, those of medicine and engineering, which are themselves becoming at the same time less dependent on tradition and more permeated by science. Its growing association with the specialized professions has tended to accentuate the separation of science from the common avocations of society. We shall have much to say in later chapters of the origin of this separation and of its dependence on the economic functions of science. It is sufficient here to draw attention to the fact that it exists most markedly in capitalist countries. Today, to many people outside its discipline, science appears to be an activity carried on by a sort of people, the scientists. The word itself is of no great age. Whewell first used the word 'scientist' in 1840 in his *Philosophy of the Inductive Sciences*: 'We need very much a name to describe a cultivator of science in general. I should incline to call him a Scientist.' These people are thought of as rather set apart: some working in obscure and inaccessible laboratories with strange apparatus, others occupied in intricate calculations and arguments, and all using languages which only their colleagues understand. This attitude has, in fact, some justification; while science grows and influences our daily lives more and more, it is not becoming more readily understandable. The actual practitioners of the several sciences have, in the course of time, moved almost imperceptibly into realms where they find it necessary to create special languages to express the new things and relations that they discovered, and have in the main not bothered to translate even the more interesting part of their work into ordinary language. Science has already acquired so many of the characters of an exclusive profession, including that of long training and apprenticeship, that it is popularly more easy to recognize a scientist than to know what science is. Indeed, an easy definition of science is *what scientists do*.

The institution of science as a collective and organized body is a new one, but it maintains a special economic character that was already there in the period when science was advanced by the separate efforts of individuals (p. 418). Science differs from all other so-called free professions in that its practice has no immediate economic value. A lawyer can plead or give a judgement, a doctor can cure, a minister can conduct a marriage or give spiritual consolation, an engineer can design a bridge or a washing machine – all things for which people are willing to pay on the spot. They are to that extent free professions in that they can demand what the market can bear. The separate productions of science, apart from certain immediate applications, are not saleable, even though in the aggregate and in a relatively short time they may, by incorporation into technique and production, bring into being more new wealth than all the other professions combined. As a result, the problem of how to live has always been the first preoccupation of the scientist, and the difficulty of solving it has in the past been the primary cause that has held up scientific advance and is still, though to a lesser degree, holding it up today (p. 1270).

In earlier times science was largely a part-time or spare-time occupation of people of wealth and leisure, or of well-off members of the older professions. The professional court astrologer was as often as not also the court physician (p. 272). This inevitably made it a virtual monopoly of the upper or middle classes. Ultimately both the tasks and rewards of science derive from social institutions and traditions, including, with more and more importance as time goes on, the institution of science itself. This is not necessarily a derogation of science. The social direction of science has been, at least until the recent drive for its militarization, a general and unexacting one, and may actually help an imaginative mind by forcing it to keep its attention on limited aspects of accessible experience. Thus, as we shall see (p. 475), the search for the longitude was a fertile social directive in the physics and astronomy of the seventeenth and eighteenth centuries, as was the search for antibiotics in the twentieth.

The real derogation of science is the frustration and perversion that arise in a society in which science is valued for what it can add to private profit and the means of destruction (pp. 844 f.). Not unnaturally, however, those scientists who see in such ends the only reason for which the society in which they live supports science, and who can imagine no other society, feel strongly and sincerely that all social direction of science is necessarily evil. They long for a return to an ideal state, which in fact never existed, where science was pursued purely for its own sake. Even G. H. Hardy's definition of pure mathematics, 'This subject has no practical use; that is to say, it cannot be used for promoting directly the destruction of human life or for accentuating the present inequalities in the distribution of wealth,' has been belied by events. For both of these results have, during and since the last war, flowed from its study. In fact at all times the individual scientist has needed to work in close connexion with three other groups of persons: his patrons, his colleagues, and his public.

The function of the patron, whether a wealthy individual, university, corporation, or a department of State, is to provide the money on which the scientist must live and which will enable him to carry on his work. The patron will in return want to have something to say in what is actually done, especially if his ultimate object is commercial advantage or military success. It will apparently be less so only if he is operating from pure benevolence or in the pursuit of prestige or advertisement; then he will only want results to be sufficiently spectacular and not too disturbing.

In a Socialist society the function of the patron is taken over by the organs of popular government at all levels, from the factory or farm laboratory to the academy institute, and is radically changed in the process. Because such a government can, and indeed necessarily must, take a long-term view, the work of scientists is accepted as intrinsically valuable. Their support and the furthering of their work are a first charge on national and local revenues. In return the scientists are expected to understand their social responsibility, which is to co-operate in the plans for a better society, and so to order their work as to get the best results on both a long-term and short-term basis.

In general the scientist has to 'sell' his project to the patron, but he is unlikely to do so unless he can count on at least the tacit support of some of his fellow scientists, through the various institutions and societies to which they belong. These bodies have the duty of maintaining the intellectual status of science, but they do not, and cannot, except where science is planned, exercise much initiative in determining the fields of science that are to be studied, nor how much or how little work is to be done in them.

In the last resort it is the people who are the ultimate judges of the meaning and value of science. Where science has been kept a mystery in the hands of a selected few, it is inevitably linked with the interests of the ruling classes and is cut off from the understanding and inspiration that arise from the needs and capacities of the people. Bishop Sprat in his *History of the Royal Society* (1667) asks himself the question: Why have 'the *Sciences of mens brains,* been subject to be far more injur'd by such vicissitudes, than the *Arts of their hands*?' He concludes that it was because they had been

banish'd, by the Philosophers themselves, out of the World. . . . Whereas if at first it had been made to converse more with the senses, and to assist familiarly in all occasions of *humane* life; it would, no doubt, have been thought needful to be preserv'd, in the most *Active*, and *ignorant* Time. It would have escap'd the fury of the Barbarous people; as well as the Arts of *Ploughing*, *Gard'ning*, *Cookery*, *making Iron and Steel*, *Fishing*, *Sailing*, and many more such necessary handicrafts have done.

If to this is added, as it has been throughout the later stages of the development of capitalism, the use of science to intensify manual work, to create unemployment, and to make war, there is an inevitable growth of suspicion and hostility to science on the part of the workers (p. 554). Science developing in this way is a limited science, hardly even a half-science, compared with its potential when it is an understood and valued part of a fully popular movement.

Any full understanding of science as an institution can come only after it has been studied from its origins in earlier institutions. It will be necessary to study the changes that it has undergone, especially in recent years, and to show how, as an institution, it interacts with others and with the general workings of society.

1.2 The Methods of Science

The institution of science is a social fact, a body of people bound together by certain organizing relations to carry out certain tasks in society. The method of science is by contrast an abstraction from these facts. There is a danger of considering it as a kind of ideal Platonic form, as if there were one proper way of finding the Truth about Nature or Man, and the scientists' only task was to find this way and abide in it. Such an absolute conception is belied by the whole history of science, with its continual development of a multiplicity of new methods. The method of science is not a fixed thing, it is a growing process. Nor can it be considered without bringing out its closer relations with the social, and particularly the class, character of science. Consequently scientific method, like science itself, defies definition. It is made up of a number of operations, some mental, some manual. Each of these, in its time, has been found useful, first in the formulation of questions that seem urgent at any stage and then in the finding, testing and using the answers to them. In the past questions to which rational answers could be given were mostly in the fields of the mathematical sciences, such as astronomy and physics. In all other fields there were only particular results found by experience and guaranteed by technical usefulness. Later, the scientific method came to be applied and modified in the fields of chemistry and biology, and now, in our own time, we are just beginning to learn how to apply it to problems of society.

Now the study of the method of science has proceeded much more slowly than the development of science itself. Scientists find out things first, and then, rather ineffectively, muse on the way in which they were discovered. Unfortunately, most of the books written about the methods of science have been by people who, though philosophically or even mathematically gifted, are not experimental scientists and strictly speaking do not know what they are talking about (p. 746).

OBSERVATION AND EXPERIMENT

The methods used by working scientists have evolved from a separation of methods used in ordinary life, particularly in the manual trades. First you have a look at the job and then you try something and see if it will work. In more learned language, we begin with *observations* and follow with *experiments*. Now everyone, whether he is a scientist or not, observes; but the important things are what to observe and how to observe them. It is in this sense that the scientist differs from the artist. The artist observes in order to transform, through his own experience and feeling, what he sees into some new and *evocative creation*. The scientist observes in order to find things and relations that are as far as possible independent of his own sentiments. This does not mean that he should have no conscious aim. Far from it; as the history of science shows, some objective, often a practical one, is almost an essential requirement for the discovery of new things. What it does mean is that in order to achieve its goal in the inhuman world, deaf to the most emotional appeal, desire must be subordinated to fact and law.

CLASSIFICATION AND MEASUREMENT

Two techniques have in time grown out of naïve observation: *classification* and *measurement*. Both are, of course, much older than conscious science, but they are now used in quite a special way. Classification has become in itself the first step towards understanding new groups of phenomena. They have to be put in order before anything can be done with them. Measurement is only one further stage of that putting in order. Counting is the ordering of one collection against another; in the last resort against the fingers. Measuring is counting the number of a standard collection that balance or line up with the quantity that is to be weighed or measured. It is measurement that links science with mathematics on the one hand, and with commercial and mechanical practice on the other. It is by measurement that numbers and forms enter science, and it is also by measurement that it is possible to indicate precisely what has to be done to reproduce given conditions and obtain a desired result (pp. 120 f., 176).

It is here that the active aspect of science comes into the picture – that characterized by the word 'experiment'. An experiment, after all, as the word indicates, is only a trial, and early experiments indeed were full-scale trials. Once measurement was introduced it was possible not only to reproduce trials accurately, but also to take the somewhat daring step of carrying them out on a small scale. It is that small-scale or model experiment that is the essential feature of modern science. By working on a small scale far more trials can be carried out at the same time and far more cheaply. Moreover, by the use of mathematics, far more valuable results can be obtained from the many small-scale experiments than from one or two elaborate and costly full-scale trials. All experiments boil down to two very simple operations: taking apart and putting together again; or, in scientific language, *analysis* and *synthesis*. Unless you can take a thing or a process to bits you can do nothing with it but observe it as an undivided whole. Unless you can put the pieces together again and make the whole thing work, there is no way of knowing whether you have introduced something new or left something out in your analysis.

APPARATUS

In order to carry out these operations, scientists have, over the course of centuries, evolved a complete set of material tools of their own – the *apparatus* of science. Now apparatus is not anything mysterious. It is simply the tools of ordinary life turned to very special purposes. The crucible is just a pot, the forceps a pair of tongs. In turn, the apparatus of the scientist often comes back into practical life in the form of useful instruments or implements. It is not very long, for instance, since the modern television set was the cathode-ray tube, a purely scientific piece of apparatus devised to measure the mass of the electron. Scientific apparatus fulfils either of two major functions: as scientific instruments, such as telescopes or microphones, it can be used to extend and make more precise our sensory perception of the world; as scientific tools, such as micro-manipulators, stills, or incubators, it can be used to extend, in a controlled way, our motor manipulation of the things around us.

LAWS, HYPOTHESES, AND THEORIES

From the results of experiments, or rather from the mixture of operation and observation that constitutes experiments, comes the whole body of scientific knowledge. But that body is not simply a list of such results. If it were, science would soon become as unwieldy and as difficult to understand as the Nature from which it started. Before these results can be of any use, and in many cases before they can even be obtained, it is necessary to tie them together, so to speak, in bundles, to group them and to relate them to each other, and this is the function of the logical part of science. The arguments of science, the use of mathematical symbols and formulae, in earlier stages merely the use of names, lead to the continuous creation of the more or less coherent edifice of scientific *laws, principles, hypotheses*, and *theories*. And that is not the end; it is here that science is continually beginning, for, arising from such hypotheses and theories, there come the practical *applications* of science. These in turn, if they work, and even more often if they do not, give rise to new observations, new experiments, and new theories. Experiment, interpretation, application, all march on together and between them make up the effective, live, and social body of science.

THE LANGUAGE OF SCIENCE

In the process of observation, experiment, and logical interpretation, there has grown up the *language*, or, rather, the languages, of science that have become in the course of time as essential to it as the material apparatus. Like the apparatus, these languages are not intrinsically strange, they derive from common usage and often come back to it again. A cycle was once *kuklos*, a wheel, but it lived many centuries as an abstract term for recurring phenomena before it came back to earth as a bicycle. The enormous convenience of making use of quite ordinary words in the forgotten language of Greece and Rome was to avoid confusion with common meanings. The Greek scientists were under the great disadvantage of not having a word – in Greek – for it. They had to express themselves in a roundabout way in plain language – to talk about the submaxillary gland as 'the acorn-like lumps under the jaw'. But these practices, though they helped the scientists to discuss more clearly and briefly, had the disadvantage of building up a series of special languages or jargons that effectively, and sometimes deliberately, kept science away from the ordinary man. This barrier, however, is by no means necessary. Scientific language is too useful to unlearn, but it can and will infiltrate into common speech once scientific ideas become as familiar adjuncts of everyday life as scientific gadgets (p. 1271).

THE STRATEGY OF SCIENCE

This discussion of the method of science has been limited so far to what might be called the *tactics* of scientific advance. This is primarily a method of solving problems and being reasonably sure that the solutions are satisfactory. It is clearly insufficient by itself to explain the advance of science as a whole over long periods of time. To complete the picture it is necessary to say something of what corresponds to the *strategy* of science. Now, of course, there is no absolute need for science to have a conscious strategy in order to advance, and indeed in earlier times it certainly was not directed with any long-term ends in view. Nevertheless, as we shall see, the path of advance of science was by no means a random one, and all the time something like a strategy must have been operating, unconsciously for the most part, but sometimes consciously as well.

The essential feature of a strategy of discovery lies in determining the sequence of choice of problems to solve. Now it is in fact very much more difficult to see a problem than to find a solution for it. The former requires imagination, the latter only ingenuity. This is the sense of Kosambi's definition of science as the *cognition of necessity*. The general advance of science has, in fact, taken place in following out the solutions of problems set in the first place by actual economic necessity, and only in the second place arising out of earlier scientific ideas. At any given time there are usually a set of challenging problems like the doubling in bulk of the cubic altar at Delphi, which involved extracting a cube root, or the finding of the longitude, which led to Newton's laws, or the curing of the silk-worm disease in France, which helped Pasteur to arrive at the idea of the germ theory of disease. The danger in science is that the number of such recognized classical problems tends to be limited. The efforts of scientists, generation after generation, are concentrated on solving them and on elaborating on the solutions.

It is this tendency that has kept science for long periods of its history within narrow bounds. It is by breaking with it and finding new problems in outside life that it expands into new fields. Some of the greatest scientists of the past, like Newton, Darwin, and Faraday, set themselves to find and solve problems according to a plan of their own. Faraday, [1.56] for instance, early in his career set himself the general problem of finding the connexions between the separate forces of physical Nature – light, heat, electricity, and magnetism – and taking them pair by pair, nearly completed the programme (p. 610).

Now we are beginning to see that what could be done consciously, though on a small scale, by such great individuals is an essential part of the growth of science, and we are finding it possible to plan science consciously on a collective rather than a purely individual basis. Here the wider problem comes from the need to reconcile and combine the questions arising from the social and economic requirements on the one hand, and the intrinsic developments of science on the other. This, however, involves, for its full advantages to be discovered and used, a far greater control over the economic life of the country than is to be found outside Socialist countries. These advantages are, nevertheless, so great in the long run that no nation will be able to hold its own in the world without making positive and planned use of science. Consequently, the advance of science and its increasing utilization in social life are likely in the future to take a far more rational and less accidental course than in the past.

Viewed in the perspective of evolutionary history, science marks a conscious elaboration of the experience provided by the sensory and motor organs of the body. It extends consciously and socially the unconscious processes of learning, common to all higher animals. An animal can learn by experience; man in using science goes beyond this and experiences to learn. In the same sense the scientific method itself, with its codified processes of comparison, classification, generalization, hypothesis, and theory, is an extension of the mechanism of the brain, which had already evolved in the higher mammals the capacity of dealing with highly complex situations, such as those involved in hunting. The essential difference, however, between these animal performances and the achievements of human science is that the latter is no longer an individual but a social achievement. It arises from the co-operative effort of *work* and is co-ordinated by *language*.

SCIENCE AND ART

The extension of the physical powers of man through science is no longer, as in animals, a continuous, almost automatic, evolutionary process. It comes about as a necessary correlate of social changes and is marked by the same internal struggles and conflicts of successively emerging classes. Bearing always in mind the inseparability of science from society, it may yet be useful to abstract still further and to consider the features of science which distinguish it from other aspects of human social activity, such as those of art or religion. The major grounds for the distinction of the scientific aspect are that it is concerned primarily with *how* to do things; that it refers to a cumulative mass of knowledge of fact and action; and that it arises first and foremost in the understanding, control, and transformation of the means of production – that is of techniques for providing human needs.

The first of these distinctions can be expressed by saying that the mode of science is *indicative*, in that it can indicate or show people how to do what they want to do. In itself the *scientific mode* does not attempt to make people want to do one thing rather than another. That is more properly the task of the *artistic mode*, a mode equally social, one of whose functions it is to generate first the wish and then the will for specific action.[1,2,146] Neither of these modes is complete without the other and, in fact, neither in science nor in art is one to be found without the other. Nor between them do they exhaust the significance of art or science for the individual. Beyond them, and common to all forms of human achievement, is the intrinsic pleasure produced in the contemplation, or still more in the creation, of new combinations of words, sounds, or colours, or in the discovery of combinations already existing in Nature. This pleasure, though felt individually in the first place, is by no means a private emotion. As the first interest derives from society, so the contemplative act is social at one remove, as is shown by the intense desire, common to artist and scientist, to communicate it.*

Every work of science has a purpose and generates a further purpose, but that purpose is not the characteristically scientific aspect of the work, neither is it the beauty or pleasure to be appreciated in the work of science. In its purely scientific aspect, it is a recipe: it tells you how to carry out certain things if you want to do them. Nor, on the other hand, is a work of art something that merely moves or pleases. Works of art themselves contain invaluable information about the world and how to live in it, especially when, as in the novel, they deal with social problems.

In stating these abstract characteristics of science, there is always the danger that the abstract may be taken as the ideal, that is, what science should be if only all the unessential aspects of social morality or usefulness could be removed. Indeed, the ideal of pure science – the pursuit of Truth for its own sake – is the conscious statement of a social attitude which has done much to hinder the development of science and has helped to put it into obscurantist and reactionary hands (p. 195). It should always be remembered that science is complete only if the indications are followed. Science is not a matter of thought alone, but of thought continually carried into practice and continually refreshed by practice (pp. 1231 f.). That is why science cannot be studied separately from technique. In the history of science we shall repeatedly see new aspects of science arising out of practice and new developments in science giving rise to new branches of practice. The professions of the modern engineer are very largely directly due to scientific progress. The very names of the different kinds of engineers there are today, electrical engineers, chemical engineers, radio engineers, indicate that they were all originally branches of science that have now become branches of practice.

SCIENTIST AND ENGINEER

But the fact that the engineers have arisen from the scientists, and are continually and closely linked with them, does not mean that the two professions are indistinguishable. In fact, the functional aspects of the scientist and the engineer are radically different. The scientist's prime business is to find out how to do things, the engineer's is to get them done. The responsibility of the engineer is much greater, in the practical sense, than that of the scientist. He cannot afford to rely so much on abstract theory; he must build on the traditions of past experience as well as try out new ideas. In certain fields of engineering, indeed, science still plays a subsidiary role to experience. Ships today, although full of modern scientific devices in their engines and controls, are still built by men who have based their experience on those of older ships, so that one may say that the building of ships, from the first dug-out canoe to the modern liner, has been one unbroken technical tradition. The strength of technical tradition is that it can never go far wrong – if it worked before, it is likely to work again; its weakness is that it cannot, so to speak, get off its own tracks. Steady and cumulative improvement of technique can be expected from engineering; but notable transformations, only when science takes a hand. As J. J. Thomson once said, 'Research in applied science leads to reforms, research in pure science leads to revolution.' [1.66.199] At the same time engineering successes, and even more engineering difficulties, furnish a continually renewed field of opportunity and problems for science. The complementary roles of science and engineering mean that both need to be studied to understand the full social effects of either.

1.3 The Cumulative Tradition of Science

So far, in discussing the institution and character of science, we have not explicitly stressed one aspect that distinguishes scientific and technical advance from all other aspects of social achievement. This feature of the sciences is their cumulative nature. The methods of the scientist would be of little avail if he had not at his disposal an immense stock of previous knowledge and experience. None of it probably is quite correct, but it is sufficiently so for the active scientist to have advanced points of departure for the work of the future. Science is an ever-growing body of knowledge built of sequences of the reflections and ideas, but even more of the experience and actions, of a great stream of thinkers and workers. To know what is known is not enough; to call himself a scientist, a man needs to add something of his own to the general stock. Science at any time is the total result of all that science has been up to that date. But that result is not a static one. Science is far more than the total assembly of known facts, laws, and theories, criticizing and often destroying as much as building. Nevertheless, the whole edifice of science never stops growing. It is permanently, as we may say, under repair; but it is always in use.

It is this cumulative nature that marks off science from the other great human institutions, such as those of religion, law, philosophy, and art. These, of course, have histories and traditions far longer than those of science, and to which far greater attention and respect are paid, but they are not in principle cumulative. Religion is concerned with the preservation of 'eternal' truth, while with art it is individual performance rather than the school that matters. The scientist, on the other hand, is always deliberately striving to change accepted truth, and his work is very soon assimilated, superseded, and lost as in individual performance. Not only the artists and poets themselves, but whole peoples look at, hear, or read the great works of art, music, and literature of the past in the original or in close reproduction or translation. They are, by virtue of their direct human appeal, always alive. By contrast, only a small minority of scientists and scientific historians, and hardly anyone else, study the great historic works of science. The results of these works are incorporated in current science, but originals are buried. It is the established relations, facts, laws, and theories, that matter for most purposes, not the manner of their discovery or presentation.[1.25; 1.32; 1.34]

There is, moreover, a profound difference of another kind between the tradition of the sciences, particularly of the natural sciences, and those of religion or of the liberal arts. The latter are arbitrary in the sense that their final court of appeal is a revelation or judgement handed down by oral or written tradition. In so far as they lay claim to a rational justification, it is one of idealist logic. On the other hand, the tradition of science, and with it that of technology, from which it arose, is one which can be directly checked by reference to verifiable and repeatable observations in the material world. However old or new, each acquisition of science can be subjected at any time to tests on determinate materials with determinate apparatus. The truth of science, as Bacon pointed out long ago (pp. 442 f.), is the success of its application to material systems, whether inanimate, as in the physical sciences; living organisms, as in the biological sciences; or human societies, as in the social sciences. It is only in so far as in the last of these there is little or no experiment that it has not yet gained the status of a true science (pp. 1019 f.). By sciences in this sense we refer necessarily to those parts of human knowledge which are sufficiently developed to be used to improve practice directly and are not merely orderly descriptions of obvious facts. It is unquestionable that the Greeks had a biology and even a sociology as well as mathematics and astronomy; but whereas the latter two could be used for the planning of towns and the prediction of heavenly events, the former only explained to the learned in an orderly fashion what was known to every farmer, fisherman, or politician. Scientific biology of real use to medicine was hardly to appear before the ninteenth century, and scientific sociology is only just beginning.

The stages by which the accumulation of scientific knowledge and techniques has taken place will be described, though not discussed in detail, in subsequent chapters. This is properly the task of a history of science, which this book does not claim to be, though such a critical history of science, going beyond the facts of discovery to ascertain the reasons, has yet to be written. Here it is sufficient to indicate some of the general principles that have ruled the building of the edifice of science.

THE PATTERN OF SCIENTIFIC AND TECHNICAL ADVANCE

In the first place, history shows a definite succession of the order in which regions of experience are brought within the ambit of science. Roughly it runs: mathematics, astronomy, mechanics, physics, chemistry, biology, sociology. The history of techniques follows an almost inverse order: social organization, hunting, domestic animals, agriculture, pottery, cooking, cloth-making, metallurgy, vehicles and navigation, architecture, machinery, engines. The reason for this is easy to see. Techniques must first arise from man's concern with his biological environment and only gradually pass to the control of inanimate forces. The actual order of the development of the sciences, on the other hand, is not so easy to explain. It is only partly conditioned by internal difficulties. In fact, as their histories show, the sciences of the more complex parts of Nature, such as biology and medicine, were derived directly by study of their subject-matter, with little help and often much hindrance from the sciences of the simpler parts, such as mechanics and physics (pp. 474, 645 f.). The time sequence of the sciences fits even more closely the possibly useful applications which were in the interest of ruling or rising classes at different times. The regulation of the calendar – a priestly function – gave rise to astronomy (pp. 121 f.); the needs of the new textile industry – the interest of the rising manufacturers of the eighteenth century – gave rise to modern chemistry (pp. 530 f.).

If we turn from the general paths of advance in science to the detailed sequences of discovery, certain general patterns appear. In any particular field there are found long chains of successive discoveries – like those, for example, of electricity in the eighteenth century (p. 599), or of atomic physics in the twentieth (pp. 729 ff.). These usually start and end with some crucial discovery that opens up whole new ranges of science. Such discoveries occur most often through the coming together of scientific disciplines previously thought to be distinct, as occurred, for example, in Oersted's accidental discovery of the effect of electricity on a magnet (p. 607), or in Pasteur's chance discovery of the asymmetric nature of molecules produced by living organisms (pp. 629 f.), which linked chemistry with bacteriology. From each of these intersections of disciplines or crucial discoveries of science there usually spring two or three new branches, each of which can continue as a new chain of discovery. The whole of the picture is therefore like an indefinitely complicated interlacing of investigation and discovery, something like the ancient Peruvian *quipu* which conveyed messages by sequences of knots on cords, themselves complexly knotted together.

THE ROLE OF GREAT MEN

Both the long chains of investigations and the branching points of crucial discoveries are essential to the progress of science, but whereas the former are the fruit, for the most part, of the application of numbers of painstaking but ordinary minds, the latter are usually associated with the great men of science. This has led to a concept of science as if it were due solely to the genius of great men, and were consequently largely divorced from the effect of social and economic factors. The hold of the 'great men' myth on the history of science has indeed lasted far longer than in social and political history. Many histories of science are, in fact, little more than the stories of great discoverers to whom came in a kind of apostolic succession epoch-making revelations of the secrets of Nature. Now great men have had decisive effects on the progress of science, but their achievement cannot be studied in isolation from their social environment. It is through failure to see this that it is so often felt necessary to explain their discoveries by resorting to 'know nothing' words such as 'inspiration' or 'genius'. Great men are thus lessened in stature and cheapened by those too limited or too lazy to understand them. The fact that they are men of their time, subject to the same formative influences and suffering the same social compulsions as other men, only enhances their importance. The greater the man, the more he is soaked in the atmosphere of his time; only thus can he get a wide enough grasp of it to be able substantially to change the pattern of knowledge and action.

Nor is the great man self-sufficient in any cultural field, least of all in science. For no discovery of any effective kind can be made without the preparatory work of hundreds of comparatively minor and unimaginative scientists. These latter accumulate, most often without understanding what they are doing, the necessary data on which the great man can work. Individual human beings have an enormous range of mental variation. Only a few are likely to contribute to science, though more have the opportunity of doing so in our time than ever before and far more are likely to do so in the near future. Those selected or selecting themselves for scientific interest are likely to differ in almost all other particulars. This gives a great variety to science, but the equally necessary unity comes from the controls, unconscious or conscious, that society exercises on it. It is this socially imposed unity of science that makes it possible to see it as one co-operative effort of man to understand and thus control his environment (pp. 1237 f.).

1.4 Science and the Means of Production

All the characteristics given in the preceding paragraphs may serve to describe science – as an institution, as a method, as a growing and increasingly organized collection of experiences. By themselves, however, they cannot explain either the major functions of science today or the reasons why science originally arose as a specialized kind of social activity. This explanation is to be found in the part that science has played in the past and plays today in every form of production. The history of the elaboration of man's means of control over his inorganic and organic environment, as it will be sketched in succeeding chapters, shows that this has taken place in stages, each marked by the appearance of some new material technique. Even now in archaeological terms (first put forward by Thomsen in 1836 but founded on traditions of great antiquity passed on by Hesiod and Lucretius) we describe the eras of the past in terms of materials – the Stone Age, the Bronze Age, the Iron Age (though we have lost the Golden Age). We continue with the ages of steam and electricity, and have now entered the atomic age or space age.

Materials in themselves, however, are no use to man; he must learn to fashion them. Even the original material (*madera* – wood – *hyle*) had to be torn from the tree to make a club or a spear. It was in the ways of extracting and fashioning materials so that they could be used as tools to satisfy the prime needs of man that first techniques and then science arose. A technique is an individually acquired and socially secured way of doing something; a science is a way of understanding how to do it in order to do it better. When we come to examine in greater detail, in later chapters, the first appearance of distinct sciences and the stages of their development it will become increasingly plain that they evolve and grow only when they are in close and living contact with the mechanism of production.

Science has had a history of remarkable unevenness; great bursts of activity are followed by long fallow periods until a new burst occurs, often in a different country. But the where and when of scientific activity are anything but accidental. Its flourishing periods are found to coincide with economic activity and technical advance. The track science has followed – from Egypt and Mesopotamia to Greece, from Islamic Spain to Renaissance Italy, thence to the Low Countries and France, and then to Scotland and England of the Industrial Revolution – is the same as that of commerce and industry. In earlier times science followed industry; now it is tending to catch up with it and lead it as its place in production becomes clearly understood. Science was learned from the wheel and the pot; it created the steam-engine and the dynamo (pp. 576 f., 613).

Between the bursts of activity there have been quiet times, sometimes periods of degeneration such as that of the later Egyptian dynasties, or of late classical times, or of the early eighteenth century. These, we shall see, coincide with periods when the organization of society was stagnant or decadent, so that production followed traditional lines and concern with it was considered to be debasing for a man of learning.

Now the observation of the close association between science and the technical change does not in itself explain the origin and growth of science; we still need to know the social factors determining the technical changes themselves. The converse relation of technical factors in society is obvious enough. The technical level of production at any period puts a limit to the possible forms of social organization. It would be useless to look for an extensive national State in the Stone Age, when food-gathering and hunting limited the effective social unit to a few hundreds ranging over a wide territory. Nor could the modern urban civilization arise until the combination of agricultural and industrial improvements made it possible to maintain a majority of the population away from the land (p. 524).

Nevertheless, changes in technique are not so simply determined by social organization. It would be very wide of the mark to assume that mankind has in the past acted as one intellectual unit, seeking always to use existing means to provide the best for all men and searching always for the best means of extending man's powers over Nature. In fact, as will be shown in the ensuing chapters, throughout the greater part of history improvements in technique have arisen mostly under the stimulus of the immediate advantage they would give to certain individuals or classes, often to the detriment of others, and sometimes, as in war – a perennial source of ingenuity – to their destruction. The form of society depends, in the last resort, on the relations between men in the production and distribution of goods – relations nearly always of undue advantage to the rich over the poor, and sometimes of direct compulsion, as in slavery.

As will be shown (Chapter 12), it is these *productive relations*, depending as they do on the technical *means of production*, that provide the need for changes in these means and thus give rise to science (p. 1072). When the productive relations are changing rapidly, as when a new class is rising into a position of power, there is a particular incentive to improvements in production which will enhance the wealth and power of this class, and science is at a premium. Once such a class is established and is still strong enough to prevent the rise of a new rival, there is an interest in keeping things as they are – techniques become traditional and science is at a discount. Such a simplified picture is, of course, inadequate by itself to explain the rise of science in detail. To discover why a certain science arose in this or that place or period requires far more detailed studies, examples of which will be found, though still only in outline, in the later chapters. It will also be necessary to bring out the interplay between the material factors – the availability of commodities such as wood or coal; the technical factors – the level and distribution of skills; and the economic factors – the supply and demand for goods or labour, in order to explain the rise and decline of science and, in turn, its effect on production.

THE CLASS CHARACTER OF EARLY SCIENCE

One basic distinction between science as such and the generalized techniques from which it arose and to which it is still attached is that it is essentially a literate profession. It is something recorded and transmitted in books and papers, as distinct from the handing on by practical example of the traditional crafts. As such, it was from the very start an occupation limited to the upper classes or to a minority of gifted individuals who managed to win acceptance into them in return for loyal service. This limitation has had several effects on the character of science. It has retarded it by keeping out of science the great majority of the naturally gifted people of all classes who might have contributed to it (pp. 553 f.). At the same time it has ensured that those thinking and even experimenting about science, at least until the time of the Industrial Revolution, should have had very little acquaintance with practical arts and so, in matters of natural science, have not known what they were talking about. Nor could they understand, because they did not themselves feel, the practical needs of common life and, therefore, they had no stimulus to satisfy them by the use of science.

This identification of science with the governing and exploiting classes has, from the earliest times of class division, which arose five thousand years ago with the first cities, engendered a deep suspicion of science, and book-learning generally, in the minds of peasants and, to a lesser degree, of the working classes. However well intentioned were the efforts of the philanthropic philosophers, the people could not but feel that in practice they would result in changes that would bring them no good, and were likely to enslave them more completely or, alternatively, throw them out of work. The first scientists were regarded as magicians capable of unlimited mischief, and this attitude persisted into late classical times when popular feeling, often allied with religion, was sullenly and sometimes violently against the philosophers who were identified, with some justice, with the interests of the upper classes (p. 233) of the hated Roman Empire. In the Middle Ages science existed only on sufferance, and even after its rebirth the same popular reaction was to be seen in the machine wreckers of the Industrial Revolution. Today we can still see it in the reactions to the latest triumph of science, the atom bomb. The combined effect of the contempt and ignorance of the learned, and of the suspicion and resentment of the lower orders, has been, through the whole course of civilization, a major hindrance to the free advance of science. It has replaced an unwilling and grudging co-operation for the free and active exchange of practical and theoretical knowledge that can, as experience in Socialist countries is now beginning to show, greatly increase the rate of technical and scientific advance.

This stricture applies only to the class character of the separation of theory and practice, and does not by any means imply any disparagement of the function of learning in advancing science. The fact that science was in the hands of persons who could write, keep accounts, and argue in set form, was at certain periods of inestimable value in its growth. Nature as a whole, taken in all its rawness and complexity, is difficult to argue about in mere words to any purpose. Myths and rituals justifying practices of proved utility are as far as such unlettered discourse can go. Even early formal science, such as that of the Greeks, was scarcely more than a rationalized mythology (p. 173). But certain parts of experience, like simple motions and forces, can be argued about formally and quantitatively. Sailors knew very well how to use oars and merchants to use balances many centuries before Archimedes discovered the formal law of the lever; but his law enabled new mechanical inventions to be made which would never have occurred to practical men. What is more, it was one step, and a very important one, in building further generalizations in mechanics and physics in the times of Galileo and of Newton. Stage by stage rational methods cease to be face-saving descriptions in a learned language and become means of generalizing and extending practical control over Nature, first in the chemical and biological, and now in the social field.

Nevertheless, as will be shown later (p. 1221), the most important and fruitful periods of scientific advance were those in which the class barrier was at least partially broken down and the practical and the learned men mixed on equal terms. Such were the conditions in early Renaissance Italy, in France of the great Revolution, in America at the end of the nineteenth century, and, in a different and more thorough sense, so they are in the new Socialist republics of today.

Just because of its universality, the class nature of science is so much taken for granted that any mention of it today in scientific circles causes a shocked surprise. The tradition of science, they feel, is something in its own right, quite separate from any considerations of economics or politics. All that this means is that the social and, in particular, the class conditioning of the scientific tradition is implicit and does not show on the surface. In our own age, for the first time, science itself is being subjected to analysis on the basis of its class character. Much of this analysis has been crude and misdirected, confusing the actual achievements of science with the general theories built into them; nevertheless, it needs

to be continued and refined and will lead in the end to a far deeper understanding of science and society.

1.5 Natural Science as a Source of Ideas

Though the practical utilization of science is both a perennial source of
scientific advance and the guarantee of its validity, the advance of science
is something more than the continual improvement of techniques. An
equally essential part of science is the theoretical framework which
links together the practical achievements of science and gives them an
ever-increasing intellectual coherence. In the past, and even now, the
history of science has often been written as if it were simply the history
of such an ideal edifice of truth. Such history can only be written by
neglecting the whole social and material components of science and thus
reducing it to inspired nonsense, as has already been stated, and will be
illustrated fully in the body of the book.

On the other hand it would be equally stupid to attempt to neglect it
entirely, for theory has had an enormously important part to play in
science, and in recent times an increasingly positive one. Indeed, over
many periods of science the main direction of work was conditioned by
the proving, or, even more often, by the disproving, of theory, as, for
example, biology in the late nineteenth century, with the proving of the
Darwinian theory of evolution (pp. 643 f.); or mechanics in the seventeenth, with the disproving of Aristotelian physics (pp. 427 ff.). There is,
however, an intrinsic danger in the development of such autonomous and
closed fields of scientific endeavour. Though starting originally from
practice they tend with time to become more and more divorced from it,
and to lose, at the same time as losing their utility, any sense of direction.
In the past they have usually petered out in learned pedantry, as did New-
tonian mechanics in the nineteenth century (pp. 541 f.); or they have been
revivified only by new contact with practice, as was electricity at the end
of the eighteenth by the discovery of the electric battery (p. 605).

The conventional view of science describes its laws and theories as
legitimate and even logical deductions from experimentally established
facts. It is doubtful, if this limitation had been strictly insisted on, whether
science would ever have existed. The laws, the hypotheses, the theories
of science have a wider bearing than the objective facts they claim to

explain. Most of them necessarily reflect in large part the general non-scientific intellectual atmosphere of the time by which the individual scientist is inevitably conditioned. The result is that the phenomena of Nature and of the manual arts are interpreted in social, political, or religious terms. Thus, as we shall see, Newton's theory of inertia came from the prevailing rational interpretation of religion, and Darwin's natural selection from the current conception of the natural justice of free competition.

Sometimes these forms of thought can lead to valid, that is practically verifiable, scientific advances. As often, especially when they win general acceptance, they are obstacles to scientific discovery. The greatest difficulty of discovery is not so much to make the necessary observations, but to break away from traditional ideas in interpreting them. From the time when Copernicus established the movement of the earth and Harvey the circulation of the blood, down to that when Einstein abolished the ether, and Planck postulated the quantum of action, the real struggle has been less to penetrate the secrets of nature than to overthrow established ideas, even though these, in their time, had helped to advance science. Nevertheless, the progress of science depends on the existence of a continuous traditional picture or working model of the universe, partly verifiable, but also partly mythical where verifications are delusive or altogether missing. It is, on the other hand, equally essential that this tradition, compounded as it is (and must always be) from elements drawn from both science and society, should be continually and often violently broken down from time to time and remade in the face of new experience in the material and social worlds.[1.6; 1.25; 1.38]*

We are passing through such a period at the present moment. The far greater part that science is playing in the economy of highly industrialized countries has, by no means accidentally, coincided with a great deepening and widening of understanding of natural phenomena, in which the discoveries of the structure of the atom and of the chemical processes in living organisms are outstanding. This in itself has put the theories of science under a severe strain, and has resulted in a rapid sequence of appearances of radically new theories, such as those of relativity and quantum mechanics (pp. 742 f.).

At the same time, and largely due to the same factors, there have occurred rapid political and economic transformations, starting in the Soviet Union and now spreading over the rest of the world, with a radically different attitude to the relations of science and society in practice. This has inevitably had a profound effect on scientific theory, which is now being subjected to a critical analysis in the light of Marxist philosophy. This will be discussed in some detail in a later chapter. As a result of these combined influences, from within and without science, there has never been a period when the theoretical foundations of science have been so much in question as they are today.

MATERIALISM AND IDEALISM

The general character of the theoretical controversy inside science is, however, not new. As will emerge clearly from a study of its history, a sometimes latent, sometimes active struggle has been going on ever since the dawn of science between two main opposing tendencies: one, formal and idealistic; the other, practical and materialistic. We shall see this conflict as the dominant one in Greek philosophy, but it must have originated much earlier, indeed from the first formation of class societies, for the general social affinities of the two sides in the conflict have never been in doubt.

The idealist side is the side of 'order', the aristocracy, and established religion; its most persuasive champion is Plato. The objective of science, in its view, is to explain why things are as they are and how impossible, as well as impious, it is to hope to change them in essentials. In Plato's mind all that is necessary is to remove a few blemishes, such as democracy, for the republic to be established safely for ever under the care of the guardians, the 'men of gold'. As the perfections of this state of affairs may not be at once apparent to inferior ranks, it is necessary to prove to them the illusoriness of the material world and consequently the unreality of evil in it (p. 194). In this imagined world, change is evil; the ideal, the good, the true, and the beautiful are eternal and beyond question; and as they are palpably not very prevalent on earth they must be sought for in a perfect heaven. This view has had a profound effect on the development of science, particularly in astronomy and physics (p. 196), and even to-day, in more elaborate and sophisticated forms, there is again a strong tendency to enforce it on science (pp. 1160 f.).

The materialist view, partly because of its practical nature and even more because of its revolutionary implications, did not for centuries find much support in literate circles and rarely formed part of official philosophy (p. 184). One expression of it, however, survives in Lucretius' Epicurean poem *De Rerum Natura* (On the nature of things), which shows both its power and its danger to established order. It is essentially a philosophy of objects and their movements, an explanation of Nature and society from below and not above. It emphasizes the inexhaustible stability of the ever-moving material world and man's power to change it by learning its rules. The classical materialists could go no further because, as we shall see, of their divorce from the manual arts; nor could, in later days, the great re-formulator of materialism, Francis Bacon. Once the Industrial Revolution was under way, science became in practice materialist, though continuing to give, for political and religious reasons, some lip service to idealism. Up to the middle of the nineteenth century materialism remained philosophically inadequate because it did not concern itself with society and its transformations, and was thus unable to account for politics and religion. The extension and transformation of materialism to include these was the work of Marx and his followers.[1.57] First effective in the political and economic field, the new dialectical materialism is only now beginning to enter the sphere of the natural sciences.

The struggle between idealist and materialist tendencies in science has been a persistent feature in its history from earliest times. The idealism of Plato is in some sense an answer to the materialism of Democritus, the founder of the atomic theory (p. 183). In the Middle Ages, Roger Bacon attacked the prevailing Platonic–Aristotelian philosophy and preached a science aimed at practical utility (p. 306), and was imprisoned for his pains. In the great struggle of the Renaissance to create modern experimental science the prime enemy was formal Aristotelianism backed by the Church. The same opposition was to be found in the last century in the warfare between science and religion over Darwinian evolution. The very persistence of the struggle, despite the successive victories won by materialist science, shows that it is not essentially a philosophic or a scientific one, but a reflection of political struggles in scientific terms. At every stage idealist philosophy has been invoked to pretend that present discontents are illusory and to justify an existing state of affairs. At every stage materialist philosophy has relied on the practical test of reality and on the necessity of change.

1.6 Interactions of Science and Society

This completes the first brief survey of the general aspects of science, as an institution, a method, and a cumulative tradition, and of the description of its links with the forces of production, and with general ideology. It should now be apparent, without pressing for a definition, what is meant by science for the purposes of this book. At the same time it would be far too much to ask the reader to accept the conclusions stated and implied in these sections without the further evidence which it is the function of the rest of the book to provide. Indeed, it is only by a fairly detailed presentation of the interactions of science and society throughout history that we can even begin to understand what science means and what its future may hold.

Science and society have, in fact, interacted in a great number of different ways, and the tendency to insist on one or other of these has been the cause of much of the recent controversy as to their mutual relations. It is usual to begin with the influence of science on society: to think of some crucial discovery, such as that of electromagnetic waves, being first of all theoretically predicted, then detected in scientific laboratories, next tried out on an engineering scale, and finally, as radio, becoming part of everyday life. But that is not the only, nor even the main way in which science grows and affects society. Even more often it happens that a scientist comes to notice the performance or the failure of some practical device. The scientist either disinterestedly, or very often with an idea of improving it, looks into it and discovers not necessarily how to make it work, but something quite different. He may indeed create a new branch of science, just as thermodynamics was founded from a study of the steam engine [1.3] (p. 585). What is important here is that common practical experience furnishes a magnet, so to speak, of scientific interest, and the progress of science can be followed in terms of successively changing fields of general economic and technical interest.

This book does not claim to be a history of science; its theme is essentially this interaction between science and society. If it has any bias, it is on the side of the influence of science on history rather than that of history on science – a theme on which much has already been written.[3.2; 1.52] But the effect of science on history has, in the past, largely been neglected, or at best dealt with in a perfunctory or misleading way. This is because the professional historians have not had, for the most part, the qualifications required to assess or even notice the contributions and influence of science; while on the other hand the historians of science have been little concerned with the wider historic consequences of the growth of natural knowledge. In official histories there has been a tendency to bring in the *state of science*, together with that of literature and art, as a kind of cultural appendage to a political, or now slightly economic, account of each historical period. What is needed instead is a discussion of the contributions of science to technique and to thought which should find its place in the very body of the narrative. To the extent that this is not done the essential *historical* character – that is the progressive and non-repetitive element – is lost from the exposition of history. We are left instead with an account of personal and institutional relations of society without any clue as to why they should not have been repeated indefinitely with variations. As clearly progressive trends cannot in fact be hidden, the non-scientific historian must blankly refuse to explain them or provide some mystical explanation, either divine providence or an assumed law of the growth and decay of civilization of the type suggested by Spengler or Toynbee. It is only in the light of science that we can begin to understand the irreversible, novelty-producing steps that are characteristically *historical* (*n.* p. 52).

As has already been indicated, and will appear in more detail in subsequent chapters, science influences history in two major ways: firstly by the changes in the methods of production that it brings about; and then by the more direct, but far less weighty, impact of its findings and ideas on the ideology of the period. It was the first of these that led to the emergence of science from technique on the one hand, and religion on the other. Once a means had been found, even in a limited sphere, of improving techniques by using organized thought ordered by logic and verified by experiment, the way was open to an indefinite influence of science in production methods. In turn these affect productive relations, and hence have an enormous influence on economic and political developments.

The other influence of science, through its ideas, was at least as early. Once formulated, scientific ideas return into the common stock of human thought. The great revolutions in man's understanding of the universe and of his place and purpose in it that have occurred in antiquity, through the Renaissance to modern times, have largely been brought about by science. It was the new reign of simple natural law inaugurated by Galileo and Newton which seemed to justify at the same time a turn to simple Deism in religion, *laisser-faire* in economics, and liberalism in politics. Darwin's natural selection, for all that it originated from such liberal ideology, was to be used in turn to justify ruthless exploitation and race subjection under the banner of the survival of the fittest. A more profound understanding of evolution was, on the contrary, to stress the way in which, through society, man could transcend the biological limits of animal evolution and achieve a more far-reaching, consciously directed, social evolution (pp. 1102 f.).

In less obvious ways scientific knowledge and scientific method are affecting, to an ever-increasing degree, the whole pattern of thought, culture, and politics. Science is now becoming a great human institution, distinct from, though closely linked with, all the older human institutions. It differs from them only in that, being more recent, it is still in its actively growing phase and its position with regard to the rest of society is not yet in equilibrium. Science has a long way to go in making its full weight felt in human affairs.

Throughout most of this book the emphasis will be on the natural rather than the social sciences, apart from the two chapters (12 and 13) devoted to them. This is because, until very recently and under the influence of Marxism, the discussion of human relations in society, itself almost the earliest of the fields of human knowledge, had not emerged from the trammels of magic and religion. In more recent times, as we shall see (pp. 1133 f.), the nascent social sciences have been reduced almost to impotence through the fear that they might be used to analyse and alter the economic and political bases of capitalism. It is partly for this reason that the social changes, brought about through the effect of the natural sciences on the mode of production, have been neither planned nor understood and have often had, indeed are still having, disastrous results. It is only by the welding of a genuine social science with natural science that a satisfactory and progressive social control of social activities can be secured.

Mankind has had at all times a 'Great Tradition', comprising the basis for what at different times has been deemed to be true belief and right action. This tradition, ever since it can be recognized as emerging from the dim past of prehistory, is essentially one, though we can discern partly independent branches in the Mediterranean countries, in India, and in China. The growth and change of this great tradition cannot be understood without science, but equally, science cannot be understood unless it is seen as a natural part of the common tradition.

The remainder of this book represents an attempt to illustrate, by a consideration of different periods and sciences, the general place of science in cultural history. According to the plan already set out in the preface, this will follow, in increasing scale and detail, the whole course of science from its first appearance to the present day. As the story is told it should be easier to understand the compressed and abstract relationships that have been set out in this chapter, and to see how they emerge naturally from the very experience of human history.

PART 2

Science in the Ancient World

.

Introduction to Part 2

Before we can understand science as we know it today, a social institution with its own tradition and its own characteristic methods, it is first of all necessary to look into its origins. Now the study of the origins of science presents a double problem. First there is the difficulty, inherent in all studies of origin, that as we go farther back and reach the critical periods where the basic innovations were made, it becomes harder to find out what actually happened. But there is an additional difficulty in the case of science, due to the fact that science does not appear in the first place in a recognizable form, but has to be gradually distinguished from more generalized aspects of the cultural life of the times. It is necessary to search for its hidden sources in the histories of human arts and institutions.

Because the essential character of natural science is its concern with the effective manipulations and transformations of matter, the main stream of science flows from the techniques of primitive man, which must be shown and imitated, not learned by rote. The expression of science is, however, initially verbal and later written, and consequently the ideas and theories of science are drawn from social life and come in turn from magic, religion, and philosophy.

The influence of the culture of ancient times affects our culture today through an unbroken chain of tradition of which only the latter part is a written tradition. The whole of our elaborate mechanical and scientific civilization has grown up from the material technique and social institutions of the distant past, in other words from the trades and customs of our ancestors. To find out about those trades and customs is the task of the historians and their colleagues – the archaeologists, anthropologists, and philologists. They must work from the material and written records of the past and from the analysis of the present customs and languages of primitive and civilized peoples.

Now in these early periods the facts are fragmentary, imperfectly known, and difficult to put together. Most are only accessible to experts in specialized fields who have usually been concerned with establishing the correct sequences and interactions of cultures and have rarely concerned themselves with the problems of seeking out the origins and influences of the sciences. Because I am neither a historian nor a scholar but a working scientist, my reconstructions are bound to be provisional and very open to criticism. It is, however, from such criticism and from the research to which it should lead, that a coherent and reasonable picture can be built up.

It would, of course, have been possible to leave out entirely a discussion of the earliest periods. A perfectly intelligible account of modern if not medieval science could be written without it. But to do so would be deceptive. So much would be taken for granted, as either self-evident or arbitrary, which was in fact the result of specific scientific and social factors operating in antiquity. For example, the great debate about the revolution of the heavenly spheres that marked the beginning of modern science is unintelligible without a knowledge of the mythological cosmological origin of these spheres, which reaches at least as far back as the earliest stages of Mesopotamian culture (p. 122).

In this section I will attempt to give in outline an account of the first creation and differentiation of science in relation to the early developments of human societies. The range of history covered falls into two major stages divided by the critical invention of agriculture. The first stage covers the period of the Old Stone Age (palaeolithic), Lower and Upper, based on food-gathering and hunting. The second stage covers the periods of primitive village agriculture (neolithic); that of the first city and river culture in Egypt, Mesopotamia, India, and China (the age of bronze); and lastly that of the independent cities based on trade (the age of iron), including the classical civilizations of Greece and Rome. It is convenient for the purposes of this book to separate out this last period, partly because it is so much better known to us from written sources, but even more because its tradition passed directly into that of modern science. Accordingly Part 2 will be divided into three chapters: Palaeolithic, Chapter 2; Neolithic and Bronze Age, Chapter 3; Iron Age and Classical, Chapter 4.

In each of these periods men made their contribution to the techniques and ideas which are the necessary basis of science. In the palaeolithic were produced all the major ways of handling and shaping materials, including the use of fire, the practical knowledge of the occurrence and habits of animals and plants in wild Nature, as well as the basic social inventions of kinship, language, ritual, music and painting. The village culture of the neolithic gave, besides agriculture, weaving, and pottery, the social inventions of pictorial symbolism and organized religion. The Bronze Age added metals, architecture, the wheel, and other mechanical devices; of even greater importance, it produced the crucial social invention of the city itself – the *civis* of civilization, the *polis* of politics. It was the city that made the technical advances possible, and with them a whole complex of intellectual, economic, and political inventions – those of numerals, writing, commerce, in the framework of a newly evolved class system and organized government. Already a conscious science was arising and the distinguishable disciplines of astronomy, medicine, and chemistry acquired their first traditions.

The Iron Age did not cause a marked transformation in material technique, though it added glass and improved tools and machines. Its chief contribution was to extend civilization far and wide by the use of the cheap new metal – iron; but the social inventions of the alphabet, money, politics, and philosophy prepared the ground for the rapid development and extension of techniques and science. It was in this period that the Greeks assembled and developed, out of the technical experience of the older empires, the first fully rational science with a direct and unforgotten connexion with our own. But the Classical period was also one of warfare and social conflict, slavery and oppression. Its final expression, the Roman Empire, gave little to science though much to public works and law. Owing to its inherent contradictions it gradually decayed politically and intellectually, and with its fall the science of classical antiquity went into eclipse though parallel branches in Persia, India, and China continued to flourish and to prepare the way for a new advance.

Chapter 2

Early Human Societies: The Old Stone Age

2.1 The Origins of Society

To find the earliest origins of science we must look into the period before there was any effective separation between the technical and ideological aspects of human culture – into the very origin of humanity itself. Now the first and most fundamental way in which human beings differ from animals is that they form continuing societies with a material culture adding new scope to the capacities of naked bodies.

Such *societies* must have had – as distinct from animal herds – better methods of getting food and protection than could be achieved by isolated individuals, and means of preserving and handing on these methods in the form of a continuous tradition. Already in their evolution from ape-like creatures, primitive men had inherited the essential bodily and mental equipment for *seeing*, *grasping*, and *handling* objects. In addition, they must have had from the outset a quite exceptional capacity for *learning* derived from a more *generalized* pattern of getting their living than most large mammals with specialized bodies and habits. It was this combined hand–eye capacity with an ability to learn[2.23] that made the use of implements possible; first the casually picked-up stone or branch, then one deliberately selected and shaped for the job. But as long as such advances were confined to individuals, however gifted, it could not constitute a full humanity. For any implement to be available to all and capable of progressive improvement, its making and use must be *taught* and *learned*. It must effectively be *standardized* by *tradition*, and that implies a continuing society.

The continuity of human societies was also made necessary and secured by the exceptionally long period during which the human infant is unable to fend for itself. This leads to a practically immortal family group through the association of different generations, particularly of the females. Grandmothers, mothers, and daughters ensure an unbroken human tradition. That is fundamentally why in primitive societies the maintenance of the tribe depends on the women. As kinship is reckoned through the mothers, such societies are called *matrilinear*. A matrilinear

stage seems to have occurred in all societies, including those of our ancestors.[2.83] It may even be that at a very early stage the women directed
the affairs of the group, so that the societies were also *matriarchal.*

Now the methods which gave human societies their particular advantages were largely dependent on the use of material *implements* for catch-
ing, collecting, transporting, and preparing food, and also on a rapid
means of communication to ensure co-operation in these tasks – in
other words, on *language*. Through the use of implements man achieves
a far greater and more generalized control over his environment than
the animal most lavishly endowed with teeth or horns. Language, by
gesture and voice, in addition to indicating the most effective use of implements, ensures both the coherence of society and the handing on of its
accumulated culture to later generations.

2.2 The Material Basis of Primitive Life

IMPLEMENTS AND TOOLS

Implements are essentially an extension of human limbs – the extension
of the fist and tooth with the stone; the arm with the stick; the hand or
mouth with the bag or basket; or an altogether new type of extension, by
projection from the body, as when a stone is thrown with aim. The social
control already necessary for the mere selection and use of implements
became even more so when such raw implements came to be deliberately
fashioned for their purpose. Every kind of instrument thus comes to be
socially determined in its use, its form, and its mode of preparation.

The continuity of tradition in primitive life is shown, right from the
beginning of the archaeological record, by the actual objects made by
primitive man himself. Even if we knew nothing about them from their
use by contemporary savage societies, they would still bear evidence of
their social origin. The implements of each type are practically identical
in any given culture or area and do not vary much over very long periods
and very large areas indeed. Now even the simplest of stone hand-axes
has to be shaped by a fairly elaborate process of chipping, a process that
would take civilized man quite a long time to learn. The fact that this shape
is preserved shows the extreme stability of technical tradition. In other
words the actual shaping of a flint implement is itself an institutional
cultural activity which has to be learnt and executed with the greatest
care in order to secure the degree of uniformity which we observe.[2.63.78]

The uniformity is not, however, absolute. There are inevitable changes:

1 a, b. Implements from the neolithic (New Stone Age) and the Bronze Age, showing development over a period of some two millennia.

improvements, borrowings, and combinations which have led, through a stage-by-stage evolution, to our present state of technique. But the important point is that through social conditioning man is able to have at his disposal at every stage of culture a reproducible, practically a standard, set of implements. Each tribal group, according to the way it gets its living, has a particular set, but many of these are common over vast areas. The habit of forming such standard sets, beginning in the early stages of primitive man, has been the major factor in preserving the absolute continuity of technical culture right down to our own time.

There is a further implication in the existence of standardized implements, namely the presence of the *idea* of an implement in the mind of the maker before setting out to make it. More than this, some partly worked flints show a definite blocking out of the material before the work is started. Later this experience of conscious foresight is to become that of *design* and *plan*, and hence of that characteristic of science – the *experimental* method. This comes from trying out various methods of making an object on models or drawings rather than always relying on full-scale trial and error.

If an implement, such as a stone picked up and thrown, is the beginning of human technical progress, that progress becomes unlimited once the *tool* is developed. The tool – the implement to make implements – creates the possibility of producing far more different types of implements than

could be simply selected or snatched from Nature. The process of making tools, first by chipping from stone, then by grinding, and finally from metal by hammering and casting, underlies all our modern techniques of dealing physically with material objects. The first stone hand-tools simply smashed what they hit; later they were developed to split, cut, scrape, and pierce.[2.53] Through the practice of tool making and tool using, men learned the mechanical properties of many natural products, and thus laid the basis of *physical* science. The use of tools not only made for far more efficient hunting; it also provided a means of shaping and preparing softer materials – wood, bone, and skin. At the same time man, or more probably woman, was beginning to put things together – pinning, sewing, tying, twisting, twining, and weaving. In this way containers for food, water, and portable objects were evolved.

CLOTHING

Partly from the need to carry things about, at first only food and implements, came the custom of attaching objects more or less permanently to the body, wherever a convenient hold could be made, in the hair, round the neck, waist, wrists, and ankles. These attachments tended to become distinctive and ornamental. Feathers, bones, and skins were added. Then came the crucial discovery that furry skins helped to keep people warm on cold nights and in winter. From this came *clothes*, first in isolated skin cloaks and skirts, then sewn and tailored garments, complete body containers, such as the Eskimos make today. These, together with skin foot protectors, enormously increased the range in area and in season of primitive man. So also, though in a lesser degree till settled agriculture came in, did wind-breaks, shelters of branches and leaves from which were to come huts and houses.

FIRE AND COOKERY

Almost every one of the early mechanical achievements of man, even to weaving and tailoring, had already been anticipated by specialized species of animals, birds, or even insects. However, one invention, the

2. The body of an iron age man preserved in a peat bog since about 1000 B.C., at Tollund, Denmark. The noose around his neck is evidence that he was hanged and, it has been suggested, for sacrificial reasons. An examination of the stomach contents has shown the presence of seeds of a kind of *Brassica* and of *Camelina*, indicating that both were cultivated at that time. *Brassica*, of course, is still with us in many forms such as the cabbage and turnip, but *Camelina* is now only found as a weed in cornfields although it was once grown for the oil extracted from its seeds.

use of *fire*, which must have come earlier than many of these, is altogether beyond the reach of any animal. How man came across fire, and why he dared to tame and feed it, is yet to be discovered. Wild fire is either confined to special localities, as in the neighbourhood of volcanoes or outlets of natural gas, or occurs very rarely, as with forest fires. The preservation and propagation of fire must at first have been frightening, hazardous and difficult, as witness the universality of fire myths and legends. At first it must have been used to warm the body on cold nights - the Australian natives carry round fire-sticks which are used instead of clothes in cold weather – and to frighten animals. Cooking could only have come once the camp fire had become an established custom.

The tool-using and fire-using animal is well on the way to a scientific humanity. Just as the tool is the basis of physical and mechanical science, so is fire the basis of chemical science. First of all came the very simple and essentially chemical practice of *cooking*. It is from this apparently almost accidental use of fire that the more specifically controllable and scientific uses of fire in pottery and later in metal-making first arose. It is not very difficult to roast meat on sticks, or even to bake roots in ashes, but boiling represents a real problem, the solution of which was to lead to further great advances. The first ingenious idea was to heat water in leather buckets or waterproofed baskets by dropping in hot stones. Such stones, cracked by heating and chilling, have been found round prehistoric camp sites. The crucial discovery, however, was that by coating a basket with thick clay it could be put on the fire and actually improved in the process. In time it was discovered, probably towards the end of the Old Stone Age, that the basket could be dispensed with and clay pottery made that would hold water and stand fire. Boiling, however, still remained a luxury, pots were heavy and not easy to carry on hunting trips. Among the plains Indians of North America the term 'boiled meat' is synonymous with a feast.

Further, once containers which could hold liquids for long periods were in use, the slower chemical changes of fermentation could be noted and used. From this new knowledge came, ultimately, the general idea of transforming materials by dipping or embedding them in reagents of which the first triumphs were the arts of the tanner and the dyer. Thus already in the Old Stone Age the set of practical recipes was built up from which rational *chemistry* was to arise.

ANIMAL LORE

The operative knowledge, however, and the use of tools and fire, is only one part, and possibly originally a rather small part, of the specific human use of accumulated and transmitted experience. Earlier, and more immediately important, was the observational knowledge of Nature, not Nature in any general sense but Nature as it appealed to man's immediate needs, principally to his need of food. The knowledge thus obtained of the habits of animals and the properties of plants formed the basis of our *biological* science of today. A very large part of the interest of primitive man must have been directed to the collection and transmission of information on plants and animals. Animals, because of their movements and of the excitement and danger of hunting them, attracted the greatest interest.

PRIMITIVE ART

For this, we have the evidence of the most detailed knowledge of Nature possessed today by all tribes still in the hunting phase and by the large part that animal dances play in their ceremonies. That this was also true in the past is shown by widely dispersed cave paintings, drawings, and sculpture, which are almost exclusively of animals. These representations do not stop at the outside of the animal, often bones, heart, and entrails are also shown giving evidence of the origin of *anatomy* arising

3. Palaeolithic (Old Stone Age) cave painting of bison from Altamira, Spain. A brush was used for details and outline, while warmed grease paints were used to fill in the remainder.

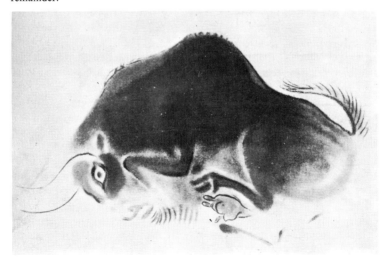

from the cutting up of game. Indeed it is to this biological aspect of primitive life that we owe the techniques of pictorial representation, which are not only the fountain of the visual arts, but also of the graphic symbolism, mathematics, and writing, which have made rational science possible.[2.90]

2.3 The Social Basis of Primitive Life

LANGUAGE

Long before such elaboration was possible, human society was evolving language, its most powerful means of cohesion and development. Language is itself a means of production, possibly the first of all. The co-operation of several individuals in the pursuit of game with their bare hands or with unshaped sticks and stones is possible only by the use of *gesture* or *words*. This may well have happened long before any instruments had been shaped for their purpose. Early language must have dealt mainly with the getting of food, including the movements of people and the making and using of implements.*

How early the acquisition of language must be is shown by the degree to which it has already influenced the inherited anatomical structure of the human brain. The complex of eye and hand co-ordination which occupies well over half the human brain is essentially only an elaboration of that inherited from an ape-like ancestor. The corresponding complex of ear and tongue co-ordination on the other hand, though not so large, is practically a new creation. It can only have arisen and have implanted itself in human heredity after the origin of society.

All mammals use their voices to some degree for social communication, but usually for that of emotion – for sex, anger, or fear – and the hearing of these cries in turn generates an appropriate emotional response. It was only later that the communication of emotion and action could add the communication of information about things and places. The transition is not complete, the undertone of emotion in language comes to the surface in poetry, song, and music, but it is never absent from spoken language and gives it a moving and even compulsive character which has contributed to the belief in the *magic* of words. Yet the magical aspect of language has always been subordinate to the utilitarian one (*n.* p. 41).[2.82]

Language must have been, from the very beginning, almost entirely arbitrary and conventional. In each separate community the meaning of sounds had to win acceptance and be fixed by tradition into a complete language capable of dealing with the totality of material and social life. For the same reason languages are as diverse as language is universal.

SYMBOLISM

The objects and situations which language is used to deal with are always far more complex than the sounds used to describe them. As a consequence the words of a language are necessarily *abstract* and generalized *symbols*. They are sufficient, and no more, to indicate the conventional action that the situation demands. In the very act of creating their languages human societies are forced to generalize, to let one word stand for many different things and to use a verbal symbolism or shorthand. The manipulation of these symbols in the brain together with direct visual *imagination* constitutes human *thought*. The *formulae* and *theories* of science are only natural and guarded extensions of the process of framing a language. Verbal symbolism, as we shall see, can be the source of error as well as of knowledge. If the emphasis is laid on the compulsive emotive aspects of words, they can become magic *spells*. If the symbol is taken for the material object or action, they may be the counters of idealist *logic*.

EARLY SOCIAL LIFE

Language, for all its variety and capacity to change, has a permanency far greater than that of technique. The Stone Age is over and done with, but the languages we use today are basically those that must have been ·spoken by some stone age tribes. So the study of language – a living relic of the past – should be an essential supplement to the study of the surviving relics of material cultures.[2.82; 2.83] Both, together with evidence from existing primitive people, should be able to provide some picture of the social life of early times. This is not the place - nor am I the person – to attempt such a picture, but only to indicate those parts that are relevant to the origin and influence of science.[2.77]

The relations of the members of a social group to each other must from the outset have profoundly modified the activity and feelings of individual men and women. The finding of food, its preparation and sharing out, the very eating of it in set and often ceremonial meals, were all social acts. They were specifically human because they mark a profound change from the unconditioned animal reaction to food – always eating it when hungry and keeping others away from it. Man's reactions on the other hand are highly conditioned through the traditional customs set up for the maintenance of the social group. To put it in another way, *man is the only fully self-training animal.* In contrast to other mammals, where instinctive training is carried out by parents for the first few days or weeks of life, every human being who comes into the world is put through an elaborate process of education, beginning at birth and lasting for many years. The process of social conditioning or *education* is strictly traditional, and the tradition has maintained its continuity and changed very slowly from the beginning of society till this day (p. 1028).

FOOD-GATHERING AND HUNTING: THE DIVISION OF LABOUR

Now the general ecological character of the human groups was determined at first almost exclusively, later very largely, by how they got their food. To begin with, they must have collected anything they could eat – seeds, nuts, fruit, roots, honey, insects, and any small animals that could be caught with the bare hands. We know nothing, except by inference, of life at this stage. All primitive peoples still surviving have passed into the next stage, where food-gathering is supplemented by hunting large animals. From the implements left behind it is possible to follow the increasingly elaborate techniques of hunting adapted for every kind of big game up to the mammoth itself.

The one unbridgeable social division carried over from the animal stage is that between the sexes. The necessarily small social groups of the early Stone Age maintained their continuity through the women, while the young men for the most part must have gone off and mated with girls from other groups to which they then attached themselves. This corresponded to an economic division in which the women collected fruits, nuts, grains, and grubbed up roots and insects, while men caught small game and fish. At that level there was little to choose between them as food-getters.

The further development of big-game hunting – a man's business – increased man's importance as a prime food-getter. It may be that this, combined with the extra strength, aggressiveness, and skill that went with it, led towards the end of the Stone Age to the dominance of men over women, such as they have for instance among the Australian hunters. Families tended to become *patrilinear* and tribal customs *patriarchal*. This trend may well have been reversed when hoe agriculture came in, enhancing the woman's importance.

TOTEMISM AND MAGIC

The very existence of the group depended on the daily collection of food, and this in turn depended on the supply of animals and plants living within the workable collecting range of a few miles, and on the ability of

4. The totem is found in many forms in primitive societies. It may sometimes be found almost unchanged, as in this photograph of a primitive totem pole, taken in Borneo by Hose and McDougal, *c.* 1920.

men and women to catch or collect it. Now only the latter depended on technique, and this necessarily changed very slowly. The numbers of animals and plants, on the other hand, varied widely and sometimes catastrophically. Man was entirely parasitic on uncontrolled Nature; what he could do by better techniques was only to deepen and widen the extent of his parasitism. He could not escape from it in reality till the invention of agriculture. Nevertheless he thought he could persuade and fool Nature to help him by methods which worked with his fellow tribesmen and with the animals he hunted. *Magic* was evolved to fill in the gaps left by the limitations of technique. By making each useful animal or plant the *totem* of a particular tribe or section of a tribe, by the use of images, symbols, and imitative dances, the primitive tribesmen believed that the animal or plant could be encouraged to flourish and multiply. This also led to food exchanges between different totem groups. Thus the elaborate social rules for relationships and for the sharing of food and ornaments could be linked together in one complex system. As long as the rules of the totems were strictly followed the reproduction of the tribe and of its food supply would be secure. Linked to the totem is the ascription of powers to certain persons, animals, or objects; they are *taboo*, *sacred*; they can only be handled in accordance with the strictest rules whose infringement brings frightful penalties. The concept of an object with latent *mana*, power, or virtue has underlain, sometimes fruitfully, the development of science. For example, the fascination of the magnet, with its virtue of attracting iron, created the science of *magnetism*. Most often, as the virtues were imaginary, the worship of objects has prevented clear thinking, as with the importance given to that useless metal – *gold*.

The totemic system is still in operation among many primitive peoples today. Traces of it are to be found in all civilizations, including our own, especially in the most conservative spheres of religion and language. Indeed, as Thomson has shown,[2.83] the whole of our terms of relationship – father, sister, uncle, etc. – are only to be understood in terms of totemic relationships. We still preserve in our lions and unicorns the relics of the totem animals transmitted through heraldry.

RITUAL AND MYTH

More immediately relevant to science are the *rituals* concerned with totem ceremonies, particularly those of birth, initiation, and burial.* That initiation rites were practised in the Old Stone Age is shown by the finding in caves of the indentations made in the soft clay by the participants of such rites and also by prints of mutilated hands. These rites, through which everyone had to pass, were accompanied with hymns expressing explanations or *myths* of the origin and development of the world in totem terms. This was the first formal *education*, that is the inculcation of a set of explicit beliefs about the world and how to control it, which completed, though it never took the place of, the practical apprenticeship of the actual techniques of hunting, cookery, etc. One of the features of initiation ceremonies was the giving of *names* which, because they implied the relation of the candidate to the totem ancestors and consequently to the whole world, were considered of special importance and sanctity. Indeed, as etymology shows (nomen – name= *gnosco – I know*), knowledge of names was the first explicit *knowledge*.[2.83; 2.3]

All myths in their first formulation must reflect the level of practical technique and social organization of their period, but, because of their association with rituals deemed necessary for the preservation of the life of the tribe and indeed of the universe, they change more slowly than the change of conditions and often become unintelligible till re-interpreted in more up-to-date terms. The myth of the Garden of Eden, for instance, originally reflected the change from hunting to agriculture, but it has been used to cover the ideas of taboo, of sex, of the wickedness of knowledge, of blind obedience to God, and of original sin. Myths, even from different tribes, blend easily and go to form a somewhat incoherent common *mythology*. It is from such totemic myths, after many changes but with an unbroken continuity of tradition, that not only the *creeds* of the religious, but also the *theories* of science have come down to us.

2.4 The Origins of Rational Science

The different kinds of knowledge acquired by primitive man – those from implements and tools, from fire, from animals and plants, and from the rituals and myths of society – were not, at their first winning, at all distinct. Wherever they existed they blended into one common *culture*. To understand the genesis of science out of such a culture it is not sufficient to describe its development in terms of the experience of the men of those times. It is also necessary to examine it in the light of modern science. We have to assess the range of what was known at any period and in any field of experience in comparison with the relative complexity of what there is to know. A fully *rational* and usable science can arise only where there is some hope of understanding enough of the inner workings

5. The creation of Eve from Adam's rib. From a woodcut in the *Margarita Philoso-phica* (*Philosophical Pearl*) by Gregory Reisch, Heidelberg, 1508. This book, discussing all manner of natural philosophy, was first published in 1496 and went through many editions. This illustration contains not only the central figures, but also makes clear the various forms of life as described in the first chapter of Genesis: '. . . the fish of the sea, . . . the fowl of the air . . . the cattle . . . all the earth . . . and every creeping thing that creepeth upon the earth.' The four basic terrestrial 'elements', earth, air, fire and water, (see page 199) are also incorporated in the picture.

of a part of the environment to be able to manipulate it at will to human advantage. Now objectively, the inanimate world is simpler than the animate and much simpler than the social, so that it was intrinsically necessary that the rational and ultimately the scientific control of the environment should follow that order.

By making and using implements, man was transforming Nature according to his deliberate will. This was the origin of rational *mechanics* – the laws of the movement of matter in bulk expressed in the practical handling of the trap, the bow, the boomerang, and the bolas. Even without such an understanding of the workings of Nature it was still possible for man to take advantage of any portion of the environment in which there was any sign of regularity. It was only necessary for man to know what to expect, without any need to bring things about himself, and to be there to take what Nature gave. This is the field of the *observational* and *descriptive* sciences, such as are the basis of the arts of hunting and of gathering fruits in their season. Beyond what might be controlled by direct human action and what might be expected from Nature, man still strove to exert his power, but by other means, at first magically, later in terms of religion.

The interests of primitive man were in any case severely limited and practical. They were confined to the provision of the necessities of life – food, animals, and plants – and the materials for tools and equipment, together with other things, such as heavenly bodies or features of the landscape, deemed to have something to do with their abundance. If the area of the rational and the expected was small, it was still a very large part of what actually interested primitive man. As society has developed, the area of effective science has enormously increased, but the field of interest has grown as fast or even faster. There is no reason to believe that primitive man felt any less secure in his world than we do in ours. He had certainly less reason to feel insecure.

MECHANICS

The beginning of the *rational* field is built into the structure of the physical universe and of the sensory–motor mechanism that had been evolved by animals in the course of thousands of millions of years in such a way that at each stage they make the best use of it. In the first place it derives directly from the visual–manual elements in the human body itself – the inherited eye–hand co-ordinations that gave man such advantage over other mammals, especially when he became a social animal. To put it another way, the possibility of rational thought for man begins in his relation to his *physical* environment. With a very simple device like a lever, for example, it is possible to know in advance what is going to happen at one end when you move the other. It was on the basis of this eye–hand co-ordination that the rational science of mechanics first grew up. It was in this field, and in the first place this field alone, that it was possible to *see* and *feel* intuitively how things worked. This was enormously reinforced by the knowledge acquired in early techniques. The roots of statics and dynamics are to be found in the shaping, making, and using of implements. Thus it was that, long before any other science could exist, man had already achieved an inner and essentially mathematical logic in the physical handling of definite and discrete objects. As science advanced it was this physical aspect that always retained the lead in rationality over other aspects of science.

CLASSIFICATION IN PRIMITIVE SCIENCE

It was only later, many thousands of years later, that the same physical methods could be used to deal with other aspects of human experience – the chemical and biological – and to make them as logically understandable and controllable. This does not mean, however, that the foundations of the biological and social sciences were not laid at this time, but only that they necessarily, from their very inner complexity, had to follow a different course. It was impossible to *see*, in the same rational way, what would happen as a consequence of any action in cookery or brewing. But it was possible to *know* what would happen first by trying and then by remembering or being taught. In this field, and even more in that of animal behaviour, knowledge was essentially traditional. It was then also strictly irrational because it was impossible with existing knowledge to understand and see the reasons *why* things happened. It did not, however, necessarily *seem* irrational, because the very familiarity of the experience made explanation unnecessary. In any case some mythical explanation could always be found, often in terms of abstract but personified operators like totem ancestors or spirits. The distinction between the rational and the descriptive fields was therefore never absolute. Further, there were plenty of likenesses and comparisons which could be made; whole classes of phenomena were roughly similar. It was in fact in this field that the practice of classification appeared which led to the development of the biological and to some extent the chemical sciences. These first classifications were necessarily embodied in language, which contains implicitly a theory of beings or things (nouns) capable of actions or passions (verbs). Here also arose a kind of descriptive reasoning by analogy, most often based on magic, which, though false at the outset, became more and more sure with the accumulation and sifting of experienced facts. Judging from the testimony of present-day savages, primitive peoples must have made a fairly clear distinction between the fields of experience in which they had reasonably good control over things; those in which they could make a good estimate as to what would happen; and those in which they had to rely on ritual and magic. Nevertheless the close interlinking of these aspects made for very stable cultures.*

THE SANCTIONS OF TRADITION

The extreme slowness of change, as is borne out by the archaeological record, shows how closely early men clung to tradition in all fields. This was possibly because they felt implicitly the unity of all their culture and the danger of straying from tradition in any part of it. How could they know that any failure to carry out the customary rituals and to say the magic words would not result in the sudden overturning of the whole order of Nature: that it would not cut off the sources of food or bring disease? It was safer not to vary anything unless circumstances made it absolutely impossible to maintain the old tradition.

2.5 The Transformation of the Environment

So far we have discussed the origins of science in primitive society only in an extremely generalized way, emphasizing how its necessary adaptive responses gave rise to an increasing and better-ordered *knowledge* of the material, biological, and human environment. But this is only one side of the picture. The other is the development and use of techniques by early man, themselves altering that environment and driving him on to further fundamental changes in the pattern of life. They did this in two ways.

In the first place each new technique enlarged the area of the usable or controllable environment. A new type of weapon, such as the bolas, for example, already fully developed in the Old Stone Age, made possible the hunting of fleet game in open plains. New equipment might have even more important consequences. Fur clothes, huts, and fire enabled early man to winter in the north. Such revolutionary technical changes enabled mankind to spread to new areas and to live in greater density in the old areas. In the second place the successful use of a new technique, such as burning a forest for clearance, would in the long run physically alter the environment itself and lead to new problems for which technical change offered the only alternative to extinction. Other crises, often indistinguishable by primitive man from those due to his own activity, were produced by uncontrollable changes in the physical environment due primarily to climatic variations. Both required either movement away from old areas or the development of new techniques to deal with the new conditions. Whether technical changes developed from within the culture or were imposed on it by changes in external conditions, they certainly occurred. Further, as the archaeological record shows, the changes were mainly progressive and gave greater control over a wider section of the environment.

EQUIPMENT AT THE END OF THE OLD STONE AGE

Already towards the end of the Old Stone Age the archaeological record shows man equipped with a rich array of technical devices – huts, sewn

6. The potter's wheel is old, and an ancient surviving portion has been dated as early as 3250 (±250) B.C. This traditional craft is still practised almost unchanged, as this photograph of pottery making in India shows.

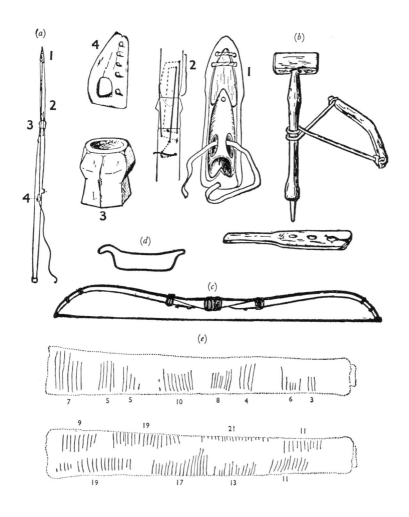

7. Primitive technology. Drawings of Eskimo equipment: (*a*) harpoon with detachable head with component parts shown enlarged on the right; (*b*) bow drill; (*c*) composite bow; (*d*) outline of stone age skin boat from Norway, similar to Eskimo *uniak*. In the lower half of the picture are two examples of early mathematics; (*e*) bone handles from Ishango in central Africa giving evidence of very early knowledge of multiplication (above) and of prime numbers (below).

fur garments, bags and buckets, canoes, hooks, and harpoons. It can be interpreted the more easily because most if not all of these are found in active use among the present-day savages, notably the Eskimos, and in a more restricted way among the South African Bushmen and the Australian aborigines. Their technique was one limited to food-gathering and hunting. Not only was the major direction and aim of life turned to the pursuit of animals, but the equipment of the hunters was very largely made from the remains of the animals they had killed. It was on the basis of such a hunting economy that solutions were found to most of the mechanical and technical problems of shaping and joining material.

It is interesting to note that although the materials have changed, most of the types of solution found at that time for these problems are still in use and are often still the main basis of modern techniques. For instance, one of the major early problems of civilization was to find means for preserving and carrying liquids about. The first buckets and bottles were of skin, and although the materials have changed, the methods of manufacture have merely been adapted to use sheet metal for buckets and cans. Even when glass and plastic have replaced leather, the essential shapes have remained the same. Basketry was certainly known in the Old Stone Age, as were crude weaving, probably derived from it,[2.90] and the plastic properties of clay. Further developments in that period of cloth and pottery were retarded not for any lack of technical ability, but because the conditions of nomadic hunting did not leave the women long enough in one place to carry out the complex operations of spinning, weaving, fulling, and dyeing, and at the same time there was little demand for goods such as clay pots, which were heavy to carry around.

MISSILES AND MACHINES

Particularly important for the history of science were the mechanical developments in hunting itself. The spear, the throwing-stick, the extremely ingenious boomerang, the sling, and the bolas, whose action depend on rather complicated dynamical and aero-dynamical movements of systems in space, are successive extensions of the simple art of throwing sticks and stones. More elaborate and far more significant for the future was the crucial invention of the bow, which seems to have occurred only in the latter part of the Old Stone Age. The bow represents the first utilization by man of mechanically stored energy, the energy stored up in bending the bow slowly in the draw being expended rapidly in loosing the arrow. The bow must be one of the first *machines* used by man, though the spring or fall trap makes use of similar principles, and may be even earlier. The bow must have led to far more efficient hunting, and its use seems to have spread very rapidly throughout the world.

For the history of science, its interest is threefold. The study of the flight of the arrow stimulated *dynamics*. The bow drill, substituting the action of the hands – and liberating one of them – in twisting a fire-stick or a borer, was an early example of sustained *rotary* motion. The twang of the bow-string was the probable origin of stringed instruments and thus contributed to the *science* as well as the art of *music*. The other and probably earlier mode of producing musical sounds was from wind instruments, of which the horn and the pipe must go back to Old Stone Age times. Primitive man knew well enough from experience that air and wind were material. *Pneumatics* started with the breath. It could be directed by blowing or sucking through hollow bones or reeds. Air could be stored in bladders for floats, and pressed out of bellows for urging the fire. Its spring could be used in the *blow gun* for hunting or in the bamboo *air pump* for producing fire. This movement of a free or driven piston in a cylinder was to be the genesis of the cannon and the steam engine.

2.6 Social Organizations and Ideas

Naturally, because the records are material, we know far more of the technical achievements of primitive man than we do of his achievements in the realm of ideas; but the few indications that we have, combined with what we know of primitive races today, show that they must also have been very considerable. In the first place, it would be impossible to carry out the complex mechanical and organizational jobs of a hunting society without a considerable capacity for intercommunication and social organization. Hunting was often on a large scale, and of such animals as the mammoth or the wild horse, involving the skilful disposition of hundreds of men.

There is, moreover, direct evidence of the development of myths and rituals in palaeolithic sites, particularly in burials. The very fact that burial was practised from almost the beginning of the Old Stone Age indicates an attitude towards the fate of man after death. The attitude seems a somewhat simple one; burials with implements and food are indications of belief in an after life not very different from that of contemporary religions. But certain practices, such as that of covering the corpses with red ochre to simulate blood, indicate a very considerable practice of magic. This is also borne out by all the remarkable paintings that lower palaeolithic man has left us in caves and rock shelters (p. 71). These paintings themselves are of an essentially magical nature and are aimed mainly at providing better hunting and more animals to hunt.

It is fair to assume, on the analogy of present primitive tribes, that this evidence points to a whole complex of ritual, essentially composed of dances and songs re-enacting the success of the hunt with masked dancers representing the animals. It is from such ceremonies that the arts of the theatre as well as the rituals of religion must descend. The imitation of animals was of course for the purpose of *deceiving* them, and its success would not long be confined to animals. The deceptive actions would be transferred to war and the poetic fiction could easily degenerate into the plain lie.

THE MEDICINE MAN

At first all must have participated in ritual ceremonies, but towards the end of the Old Stone Age there is evidence of some beginning of specialization. The paintings in remote and inaccessible caves must have been carried out by trained artists who must, moreover, have still participated enough in the chase to find and study their models in action. Among these paintings are occasionally found single figures of men dressed as animals who seem to have had some special importance. In most primitive tribes today we find medicine men or *shamans* who are thought to have peculiar relations with the forces deemed to control those parts of the universe or environment that matter – primarily food, but also health and personal luck. These people are, to some degree, set apart from the whole-time work of food and implement production, and in return they exercise their magical arts for the common good. They are also responsible for the conscious preservation of traditional learning and consequently for its modification in a developing society. Their forerunners in ancient times are therefore the lineal cultural ancestors of sacred kings, priests, philosophers, and scientists.*

THE THEORY OF MAGIC : SPIRITS

The operations of the magicians were based, probably at first only unconsciously but afterwards explicitly, on an essentially imitative and sympathetic kind of theory of the working of the universe. From the

8. Drawing of a painting from the caves of the Trois Frères in the Ariège district of southern France. Similar in age to the bison illustrated in plate 3, it depicts a sorcerer wearing stag's horns, an owl mask, wolf's ears, the forelegs of a bear and a horse's tail. This may have been a means of ensuring successful hunting.

evidence of the burials and pictures it would seem that this was already elaborated in the Old Stone Age. First likenesses and then simplified *images* or *symbols* could be so identified with the originals that operations on them were transferable by sympathy to the real world. An unbroken sequence links those images and symbols to those we use with such success in modern science, but centuries of experience and bitter struggles were necessary to distinguish the magical from the merely conventional value of symbolism.

Another aspect of primitive thought which at some stage split off from imitative or symbolic magic was the idea of the influence exerted in the real world by *spirits*, and consequently the need to control or propitiate them. The idea of a spirit is in itself a highly sophisticated one. It probably originated from the inability to accept the fact of death; and early spirits, as the grave furnishings show, were conceived as very corporeal indeed. But because they had been members of the tribe when alive they were deemed to continue their concern with it. They were thought to work on Nature as live men did by direct action or by magic, and originally their power was no greater. It was only later that the spirit (breath, geist, soul, psyche) – that which left the body at death – was imagined as separate from the body and capable of an invisible but not ineffective life of its own.

Ultimately, the conception of spirit was to split into two very different ones. The first was the transformation of the spirit of a powerful man through that of a legendary *hero* into that of a *god*,[2.73; 2.83] to become the central figure in *religion*. The second was the divorce of the spirit from its human origin into an invisible natural agent such as the wind or the presumed active force behind chemical and vital changes. The latter, purged of its divine nature as will be shown in subsequent chapters, played an enormously important role in science, ending up when condensed as the 'spirits' of the gin-shop, or remaining as the 'wild untameable spirits' – the gas (or chaos) of van Helmont (p. 620) – that were ultimately to pass into the confinement of the gasometer.

2.7 The Achievement of Primitive Man

This all too brief survey of the techniques and ideas of primitive man should at least suffice to show how much had already been done by the end of the Old Stone Age in using human intelligence to control Nature by material instruments, and, through the workings of society in tradition and ritual, to ensure that the advances gained should be retained. The basis of *mechanics* and *physics* had been established in the making and use of implements, the basis of *chemistry* in the use of fire, and that of *biology* in the practical and transmissible knowledge of animals and plants. Social knowledge was implicit in language and the arts, and had been systematized in totemism with the beginning of formal education in initiation ceremonies.

The character of the society, determined by its dependence on hunting and food-gathering, was essentially communal, without any marked specialization and without class divisions.

THE LIMITATIONS OF A HUNTING ECONOMY

The excellence of the technical and social achievements of palaeolithic men was such that one may wonder why they were not able to maintain themselves indefinitely in that state. Indeed, some have apparently done so, but only in the most outlying places, such as the Arctic and central Australia, or in the tropical forests. It is still, however, doubtful how far these are really palaeolithic survivals or merely neolithic groups pushed back by especially hard external conditions on to a secondary palaeolithic hunting and food-gathering culture. For the rest, palaeolithic technique was perhaps too well adapted to its main purpose of hunting a limited number of species of game in a limited number of habitats, particularly open plains. If the conditions determining their abundance were altered, either by climatic changes or by over-hunting by the tribes-men themselves, the herds would die off and the tribesmen would either have to move off to more favourable regions, die away on the spot, as many tribes did and are doing today, or learn to change their hunting culture for another – a far more difficult task.

The essential weakness of a hunting society is that it is parasitic on the animals it hunts. It is able to make the greatest use of the animals that are there, but not to control them in any positive way; that is, it can kill off the animals, but it cannot feed them or make them breed. In fact it was probably the very efficiency of late palaeolithic techniques that caused the disappearance of large animals from wherever they could be easily hunted. Another contributory cause was changes of climate, which replaced the open happy hunting grounds by forests in some regions like western Europe or by deserts in others, as in Africa. Certainly hunting, about the period of the end of the Ice Age, ceased to be the most pro-gressive type of human culture, and though its arts and even its social organization were preserved, they persisted only as a part of a far richer and more progressive culture brought about by the invention of agriculture.

There may also have been internal reasons rooted in the form of palaeolithic society that made it less able to cope with its environment, but it is still difficult to analyse them. Primitive societies of this level of material culture are rare today, and their purely internal difficulties are masked by the destructive influence of more advanced cultures, particu-larly our own.

Chapter 3

Agriculture and Civilization

3.1 Towards a Productive Economy

This chapter covers the periods usually known as the neolithic or New Stone Age and also the Bronze Age – the period of the early river civilizations of Egypt, Mesopotamia, India, and China. No attempt will be made to trace the histories of these civilizations, but only to bring out the part they played in the origins of science.

About 10,000 years ago there began a revolution in food production that was to alter the whole material and social mode of existence of man. This was mainly, if not altogether, a result of the crisis in hunting economy discussed at the end of the previous chapter. The difficulties that men had to face at that time led to an intensive search for new or even old and despised kinds of food, such as roots and the seeds of wild grasses. This pursuit was to lead to the invention of the technique of *agriculture*, ranking with the utilization of *fire* and of *power* as one of the three most momentous inventions in human history. Like all great transformations it was not a single act, but a step-by-step accumulation of interlocked inventions all subservient to the essential achievement – the cultivation of seed-giving grasses. In essence this was a transformation of society from the exploitation of the animate environment to its control, the first step in the achievement of a fully productive economy.

THE ORIGIN OF AGRICULTURE

The precise origin of agriculture is and will probably long remain conjectural. The limitation of the plants and animals used in agriculture to a very few closely related kinds – edible seed grasses, horned cattle – points to it having arisen in a definite period in some limited area, probably in the Middle East. It is not even certain whether the growing of crops and the domestication of animals were always associated or were the results of the coming together of purely agricultural and purely pastoral cultures. The evidence [2.17.75] seems to point to the former alternative. Originally animals may have been attracted by the extra fodder left by the grain-growers, and tamed. Domestication was not absolutely

9. Hoe, with iron blade, in use on hard ground in northern Nigeria. This design is of considerable antiquity but is an effective hand tool considering the nature of the ground to be worked.

new: already in the Old Stone Age the dog had been tamed. One small clue that has struck me is that the almost universal means of cutting grain – the *sickle* – is clearly, from its shape and the teeth with which it was originally furnished, a substitute for the jaw of a sheep or other ruminant which is a very effective grass-cutter.* It would hardly have been used, however, if sheep had not been fairly plentiful and presumably tameable in the very first stages of agriculture. The growing of crops is in any case a more far-reaching invention than the domestication of animals, for without supplies of fodder it is usually impossible to keep an adequate number of animals in a restricted area. Further, the market for meat, skins, and wool provided by the townsmen is essential to an extensive pastoral economy. A nomadic tribe of shepherds or cattle-men on the open ranges needs as much land as if it were hunting the same animals wild, while without a market from which weapons, ornaments, and supplementary food can be got there would be little incentive to exchange the excitement of hunting animals for the trouble of herding them.

The cultivation of grain may, however, have arisen without any violent break in culture in a well-stocked region where wild grain was abundant enough to be plucked by the women and kept in baskets in permanent settlements.† Enough seeds would get scattered around to produce crops worth reaping. The *invention* of agriculture is probably little more than a sufficiently clear understanding of this accidental occurrence to justify the practice of sowing grain as a *deliberate sacrifice* of good food, for a more ample return in the next season. This implies a certain fixity of settlement which may have been determined in any case by the limitation of open land in a forest, or watered land in a desert. There is some evidence that agriculture may have started on the alluvial fans of mountain streams on the edge of desert plains, which would be a natural point of retreat for game and men as the plains dried out.

As grain-gathering was women's business, agriculture was probably a women's *invention*, and in any case was women's work, at least till the invention of the ox-drawn hoe or plough, for it was done with the *hoe*, a derivative of the old stone age *digging stick* with which women used to grub for roots. Where agriculture predominated over hunting in providing food it accordingly raised the status of women and halted and reversed the tendency to change the reckoning of descent through the mother (*matrilinear*), to that through the father (*patrilinear*), which hunting had first induced. Only where stock-raising predominated, as in the lands bordering the agricultural settlements, was there a complete transition to the patriarchy – as we see it in the Bible.

Whatever its origin, agriculture led to an essentially new relation between man and Nature. Man ceased to be parasitic on animals and plants once he could grow in a small area as much food as he could hunt or collect over a wide range of country. In practising agriculture he *controlled* animate Nature through a knowledge of its laws of reproduction, and thus achieved a new and far greater independence of external conditions. The first agriculture may well have been a mere scratching of the ground, or garden culture, carried out in patches temporarily cleared and then abandoned, a kind of nomadic agriculture which is still practised by many tribes today. But even at this low level the practice of agriculture had an explosive effect on human material and social culture. Compared with any of the changes that occurred in the Old Stone Age, it marked a new order of advance. It led to a new kind of society which was *qualitatively* different, because of the enormous *quantitative* increase in the number of people that could be supported on the same land. Hunting had to

10. Sowing using ox-drawn plough and reaping with a sickle. The plough is driven by Nesitanebtashru, who is also performing the reaping. Nesitanebtâshru was the daughter of Panietchem II, priest-king Amen-Rā, twenty-first dynasty (*c.* 1000 B.C.) and was herself a priestess. This photograph is from a small section of the Papyrus of Nesitanebtâshru buried with her in her tomb, and which contains chapters from the Theban *Book of the Dead*, hymns to Osiris, to Rā, Thoth and other great gods. The papyrus was presented to the Trustees of the British Museum in 1910 by Mrs Mary Greenfield.

be a fairly continuous occupation, but agriculture depended on the seasons. Most of the population could be set free for other tasks for some part of the year. Thus agriculture brought new possibilities and with them new problems.

CRAFTS OF THE FIELD AND THE HOME

Agriculture itself involved a set of new techniques in the growing of crops and the preparing of food from them, such as sowing, hoeing, reaping, threshing, storing, grinding, baking, and brewing. With them came a whole set of ancillary techniques, either, like weaving, made possible by ample supplies of wool and flax, or, like pottery and hut-building, arising from the possibilities and needs of permanent occupation. Hut-building was known in the Old Stone Age, but only in localities where there was enough game to allow of permanent settlement. In agricultural communities it was universal. Everything conspired to put a new tempo into cultural development. The need and the material means were there. The tyranny of old customs had to yield to new conditions. One new factor was the emergence of real property, though first of *communal* and not of *private* property. In hunting communities most of what was produced was consumed on the spot, and the only permanent goods – hunting gear, cooking utensils, and clothes – were in constant and largely personal use. In an agricultural community, on the other hand, land, cattle, huts, and stores of grain were always there as more or less fixed *goods*, largely communally held, and means had to be found of safeguarding and distributing them. At first this was done by extending and further complicating the totemic group organization. The rule was share and share alike inside each group, and ritual exchanges, minutely regulated by custom, were made between groups on ceremonial occasions such as weddings and funerals. But the new methods of production were in the end to be too much for the old system of distribution. Barter began to take the place of ritual exchange, individuals began to stress their claims to what they had produced, and *private property*, with its inevitable consequence of *inequalities of wealth*, came into being. The next stage – the formation of *social classes* – does not, however, seem to have developed until the founding of cities.

WORK

Agriculture also introduced a new concept into social life: the concept of *work*. In the days of a hunting culture, work was not conceived as distinct from other aspects of life. Actions were closely related to their consequences. You hunted for food which you and yours were going to eat fairly soon. But in agriculture there was a long interval between what you did and what you got for it, and at the same time many of its operations were tedious and exhausting in themselves and lacked the excitement of hunting. True, the food supply was more secure, but the possibilities of wonderful hunts and great feasts were lost. In fact, the transition from hunting to agriculture was a transition which we now know in our legends as 'the fall of man'. Man left 'paradise' or 'eden', which means the plain or happy hunting ground, to take up working for his bread by the sweat of his brow.

SCIENCE AND THE NEW CRAFTS

Nevertheless, the very indirect relation of work to its reward that agriculture introduced led to a further extension of the concept of cause and effect which was to become the basis of a rational and conscious science. For example, the whole life-history of animals and plants now came under interested observation. It was necessary to know how they bred and grew, not only how to catch the one or gather the other. Similarly the new techniques that came in with agriculture introduced new mathematical and mechanical concepts. *Weaving* is clearly a further adaptation of basket-making, and both of them involve regularities, first of all actually practised and then thought about, which are at the basis of *geometry* and *arithmetic*.[2.90] The *forms* of *patterns* produced in weaving and the *number* of threads involved in producing them are essentially of a geometrical nature, leading to a deeper understanding of the relations between *form* and *number*. *Spinning* was, with the possible exception of the bow drill, the first industrial operation involving *rotation* and may well in turn have led to the use of the wheel, which in the next period was to *revol*utionize mechanics, industry, and transport. *Pottery*-making, on the other hand, was the first indirect application of fire and demanded far greater control of it than did lighting, warming, and cooking. The use of pottery was in turn to extend the range of cooking operations and was to make the smelting of metals and early chemistry possible.

THE NEOLITHIC AGE*

The period between the first invention of agriculture and the founding of cities is usually known as the New Stone Age or neolithic age. It was so called because of the use of ground and polished stone implements in place of the chipped instruments of the Old Stone Age. In the centres of ancient civilization it lasted roughly from 8000 B.C. to 3000 B.C. The culture characterized by polished implements covers, however, a much longer period of time, and indeed there are many peoples in the world

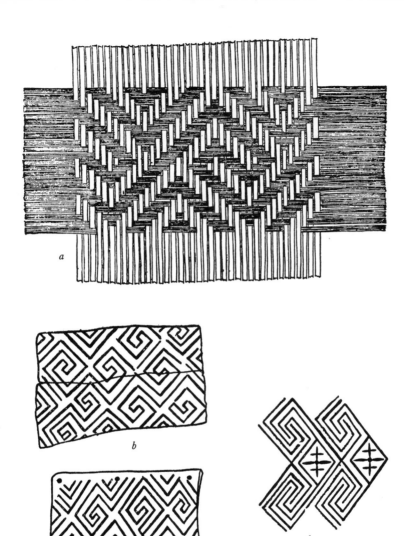

11. Basketry was certainly known from palaeolithic times and from it weaving was probably derived. Both techniques involve regularities and may perhaps have exerted some influence on men's recognition of patterns, their use in art and their later development into the geometrical figure and its mathematical analysis. Such regularities may well have led to a deeper understanding of number as well as of form. (*a*) a piece of woven matting to show how the Greek key design originates from simple alternations; (*b*) and (*c*) pieces of palaeolithic ivory from near Kiev (note mistakes and distortions); (*d*) design from tomb of Thutmose III, *c.* 1500 B.C.

today living in a state of neolithic culture. It would appear that these existing neolithic cultures have arisen in two ways. Some may be in direct continuity with primitive neolithic culture which spread widely from the original centres in the Middle East; others may have derived from a much later spread of bronze age peoples who, moving into regions where they were cut off from the products of their parent cities, lost all but their basic neolithic material culture and retained only certain bronze age ideas, such as sun-worship. The first megalithic long-barrow folk, who came to Britain 4,000 years ago, may have been such a group. So also may the Polynesians who spread over the Pacific during our Middle Ages.

The very persistence of neolithic culture over most of its area shows how in it man had achieved a new equilibrium, though now with the produce of soil and climate rather than as before with animals and plants in the state of Nature.

THE FORMALIZATION OF RELIGION

This transformation in the material basis of common life which came with the invention of agriculture was bound to have a profound effect on the mental sphere which was expressed in new rituals and myths. The chief concern of the neolithic community was with crops. Accordingly the woman's side of the totemic rituals for increase and reproduction of plants was emphasized and further developed. The most characteristic were *fertility* rites, in which human matings were used to encourage the crops. The influence of rain on vegetation, noticed in the days of hunting culture only indirectly through its effects on animal life, now became a matter of life and death. Imitative magic to produce rain became the other main object of ritual.

This concentration tended to make ritual and magic more orderly and to bring about their transformation into *government* and *religion*. Regular spring and harvest festivals were celebrated. Corn queens or kings and rain-makers were chosen and given special consideration and powers because they were regarded as essential to the life of the community. The need to bury or kill the grain before a new crop could grow led to the idea of *sacrifice*, even of a human sacrifice, in which the king himself or his representative was called on to die for the welfare of the people.

VILLAGE CULTURE

The characteristic economic and cultural unit of the neolithic age is the *village*. Now many centuries must have been needed to evolve the complex interrelationships of technical and economic operations carried on

12. Mithras slaying the bull. This Roman sculpture, now in the British Museum, commemorates an ancient Zoroastrian fertility rite, for here the bull's blood was identified with the new corn. In Zoroastrian doctrine Mithras was one of the good powers, closely associated with light and often recognized as a Sun god.

in a village that ensures its practical independence in its own territory. Village economy, however, is strictly limited in scope and possibility of change. Even where it involves thousands of people, as in some African villages today, it remains an economy in which nearly all the people are occupied most of the time in agricultural pursuits or in the production of locally made and locally used goods. The self-sufficiency of the neolithic village favoured its spread, but it hindered its further development.

3.2 Civilization

RIVER CULTURE

The first step towards a larger scale of operations occurred when people tried to practise agriculture in the wide alluvial valleys of such great rivers as were free from unclearable forest, that is, flowed in their lower courses through arid lands. They may have started from the low river banks, where seeds could be sown in the wet mud, as tribes in the Upper Nile still do, and then gradually cut back the marshes and cleared the river channels. Alternatively, the practice of agriculture in small upland valleys may simply have been pushed downstream step by step into the great valleys. In either case there would be an incentive to canal cutting and embanking. In some such way a new kind of agriculture based first on natural, then on artificial irrigation came into being. In such a territory the village ceases to be the natural economic unit. Floods and droughts do not respect village boundaries; embankments have to be raised and canals have to be cut by many villages working together, and the distribution of the water must be fairly partitioned between them. When such co-operation, even that between half a dozen villages, was achieved or imposed the yield of the land of each of them increased. This

13. The annual inundation of the land surrounding the Nile is indispensable to the agricultural well-being of the neighbouring countryside. In Egypt, without artificial irrigation, only along the banks of the river can anything be grown – in other words, over an area no more than 4 per cent of that of the total country and in which about 96 per cent of the population lives even now.

marked another quantitative advance in food production, as it enabled a still larger number of people to live on the land, and this in turn led to a qualitative change in social organization.

EXTENSION OF SOCIAL CO-ORDINATION

Social co-ordination over a far larger area than the simple village was in fact necessary to get the full value out of river-valley agriculture; but once it was achieved it was consolidated by its very success. Simply to increase the scale of an operation often leads to altogether unsuspected possibilities. When the tribes of the Nile villages federated, or were conquered so as to form one economic unit, they were able almost at once to produce so much more surplus wealth that in the space of two or three centuries they were in a position to support the enormous economic burden of the State works of the first Egyptian empire.

Another example, from more recent times, shows how important is the effect of organization by itself without notable technical changes. The Inca empire of Peru arose out of the welding together of a number of independent tribes each cultivating its own bit of valley, arranging its own limited irrigation canals, and living off its own produce. The energetic and domineering tribe of the Incas, later to become a kind of sacred aristocracy, partly by political genius, partly by force, made these tribes federate. They thus made it possible to treat whole valleys as one unit, to drive long canals, to terrace whole mountain-sides, and to arrange for the suitable division and appropriation of food. As a result, for the centuries during which their empire lasted, no one in Peru needed to go hungry. Now the interesting thing is that this system, although it did not employ any new techniques, provided a surplus of products great enough to maintain the Inca ruling classes – the children of the sun – in very considerable splendour, and it also enabled them to create within a matter of a few centuries a quite high level of intellectual culture and a remarkable architecture.

Civilization could only have originated and first taken root in the well-watered river valleys, where cultivation by natural flow irrigation canals could be practised. Later it was to spread locally by the much heavier engineering works of lifting water for high-level channels, digging wells, and terracing hillsides, but until the Iron Age it could never get far from the alluvial plains. Early civilizations were accordingly limited to a number of favoured areas, the main ones known to us being those of Mesopotamia, of Egypt, and of the Indus valleys, and, some centuries later, of the valleys of the Oxus and Yaxartes, the Yellow River, and the Yangtze (Map 1, p. 238).

14. The Inca civilization produced magnificent buildings and cut terraces into the mountainous surroundings (see page 101). Terraces in the Urubamba valley, Peru.

THE ORIGIN OF THE CITY

We are apt to think of *civilization* as arising primarily from the *city* – the *civitas* – which gives it its name. But the city was actually a consequence and not the cause of civilization. A city differs from a village by the fact that most of the inhabitants are not food producers working on the land, but administrators, craftsmen, traders, and labourers. Before a city can be founded the level of technique of agriculture must be so raised that the non-producers in the city can be maintained on its surplus. As we have seen, such an agricultural technique requires at the outset some central organization. This implies a body of administrators covering a number of villages. One of these, containing the temple of the leading totem god, would naturally become the *city* where the surplus from the remaining villages would be collected and stored. As we do not yet know where the first cities were, the transition from village to city seems more abrupt than it probably was. Of existing cities Jericho seems the oldest, for there masonry walls are found in a period so early that pottery was not known.[2.44] In lower Mesopotamia it is possible to trace a transition

between villages and small cities built on the same site. Any later foundations of cities [2.83] are bound to have been influenced by the idea or even
the experience of what a city should be like. Some evidence suggests that
cities were founded by bringing together part or all of the populations of
several villages. The site of the city itself may have been a strengthened
natural hillock, a refuge against floods afterwards sanctified as a temple
platform on which the temple stood like a mountain, the prototype of the
tower of Babel.

A city may have arisen in the first place from the village of the chief
water magician of the district, through whose instruction the irrigation
was organized. This does not necessarily imply any great innovations or

15. A street in the ancient city of Mohenjo-dāro in the Indus basin, west Pakistan.
The city flourished in the third millennium B.C., and the street shown here is 30 feet
wide; it was one of those that divided the city into blocks. On the right is a drain that
was originally brick covered.

even much conscious use of science. The digging of canals and working of sluices need at first have been little more than clearing out existing watercourses and breaking holes in naturally formed banks, very much as in historic times the elaborate dike systems of Holland were evolved from sand-pits and mud-banks. Here, as in all beginnings, art (*techne*) follows Nature (*physe*) as Theophrastus says: '. . . it is manifest that Art imitates Nature, and sometimes produces very peculiar things.'[2.80.139] To succeed, however, without confusion, the work of irrigation would need some authoritative direction delegated or assumed with religious sanctions.

Once a city was established, however, a new division appeared: that between town and country. This did not happen all at once; for centuries most citizens owned and worked lands outside the walls. The surplus provided by the new efficiency of agriculture went to the city; not much was left for the villagers to enjoy. The Egyptian peasant of early dynastic times was probably rather worse off than his ancestors in the neolithic age as regards freedom and conditions of work, though he had a better and more regular food supply. But he was no worse off on either count than his descendants, the modern fellahin.

THE EVOLUTION OF THE HOUSE

At first the cities hardly differed from villages: just an assemblage of huts, each with a courtyard for animals, the dwelling-place of one family, but usually one of several generations, together with servants and slaves. As the population grew, more huts were added to the court, often as lean-tos inside the wall, making the first real houses. These came to be made of mud brick, as the danger of spreading fires from reed huts was too great. The life of the house centred round the court; the outer walls were windowless. In hot weather the family slept on the roof under an awning and later upper storeys with windows appeared. The spaces between the houses gradually shrank into streets, though some were left for markets and the remainder for gardens. Round it all, as property grew and war threatened, was built a wall constricting and crowding it still more. When civil strife threatened as well, an inner fortress or citadel was built, from which armed men could dominate the city or into which they could retreat at need.

TEMPLES, GODS, AND PRIESTS

The city was centred round a *temple* or big house, in which one god assisted by his priests superseded or ruled over a small pantheon of local village totem ancestors.

16. A cylinder seal from Ur, third dynasty, *c.* 2050 B.C., depicting two goddesses introducing Hashamer, governor of the city of Ishkun-Sin, to the seated King Ur-Nammu, founder of the third dynasty of Ur. Now in the British Museum.

The institution of *gods* is essentially one derived from city life, and was brought about by the exaltation of the simple clan spirits through the newly available wealth. For that reason the god might well be an animal, as in Egypt, or have animal doubles like Zeus and his eagle. The first gods, as we meet them in Sumerian legends of 5,000 years ago, were very human indeed. They had their councils, quarrels and debates, very like an assembly of village elders.[2.29; 2.76] In each city sooner or later one God and his consort usually came to dominate, but the others were not abolished but assigned subsidiary roles. At the same time the growth of cities was marked by the increasing separation of the God from tribal and village concerns and His identification physically with His house in the city, and with the administration of His lands and property by His priests. From the beginning these priests ran the cities and drew the largest share of their benefits. They were the heirs of the medicine men of the Old Stone Age and of the magic kings of early agricultural communities, though in Egypt the magic king remained as Pharaoh, ruler and high priest. The *priests* formed the first administrative class, having definite and indeed essential functions; they arranged for the distribution of water and seed, for the timing of sowing and harvest, for storing of grain, and for collecting and apportioning the herds and their produce.*

TEMPLE SERVANTS AND CRAFTSMEN

The physical work needed to maintain the organization of the economy was not, however, done by the priests, or done only in a symbolic way.

We see for instance pictures of the priest-kings of the ancient Sumerian cities carrying the first basket of earth from the excavation of a canal, and of Egyptian Pharaohs wielding a hoe, much in the same way as their successors today lay foundation stones. A body of temple servants was required for collecting, storing, and guarding the surplus produce. The temple itself became an establishment needing building and upkeep and the preparation of increasingly lavish ceremonials and feasts. The table of the god had to be well supplied. The exalted god naturally appreciated only the spiritual essence of the food, while the priests had to be content with its material remains. All these activities required workers who tended to become more specialized and gradually altogether removed from the work of agriculture. Builders and carpenters, potters and weavers, butchers, bakers, and brewers congregated round the temple and shared, though modestly, in its revenues. The first complete *division of labour* took place as these craftsmen were attached to their tasks and divorced from the land. Nothing could be too good for the gods, and, with supplies of materials assured from the agricultural surplus, the craftsmen rapidly improved their techniques. New crafts such as that of the jewellers and metal-workers were added to the old. In the cities the old clan organization of the villages, already overstrained by the appearance of property, was reduced to a formal role or continued as a guild mystery for the followers of particular crafts.[2.83.332]

CLASS-DIVIDED SOCIETIES: SLAVES AND SERFS

Up to the present little has been written on the original process of transformation of village into city economy. The evidence may be there, but it has not been fully interpreted. We badly need an economic and social

17. Diorite stele 245 cm. high on which are carved the laws of King Hammurabi. Hammurabi ruled *c.* 1800 B.C. and the stele was originally erected at Sippar (Babylonia). It predates the Mosaic laws by about a century and a half and is the earliest legal code that has been discovered in a sufficiently complete condition. At the top of the stele a low-relief carving depicts the Sun god (Shamash) offering the code of laws to Hammurabi and below is the code. Divided into 282 articles and preceded by an invocation in which the greatness of the king and the excellence of his purpose is extolled, the code deals with landed property, movable property, business matters, the family, injuries, and laws governing labour. The code was imitated and partly followed not only by the Hebrews but also by the Hittites and the Assyrians. Part of the inscription (on the opposite side of the stele shown here) runs: 'If a man destroy the eye of another they shall destroy his. If one break a man's bone they shall break his bone. If one destroy the eye of a freeman or break the bone of a freeman, he shall pay one mana of silver. If one destroy the eye of a man's slave or break the bone of a man's slave he shall pay half his price.' The stele is now in the Louvre.

analysis of the really primitive bronze age cities, similar to that which Thomson has given us of the iron age cities of Greece.[2.83] When the archaeologist reveals them the earliest cities seem far along the road of *class-divided societies*, and early laws are quite explicit on the point. In the Code of Hammurabi (*c.* 1800 B.C.) for instance we find in a table of retributory penalties the following:

If a man destroy the eye of another man they shall destroy his.

If one break a man's bone they shall break his bone.

If one destroy the eye of a freeman or break the bone of a freeman, he shall pay one mana of silver.

If one destroy the eye of a man's slave or break the bone of a man's slave he shall pay half his price.[2.35]

This implies three grades. In most early cities we find citizens graded according to their wealth, including priests, merchants, and free craftsmen; there are domestic slaves and, outside the city, there are peasants who are virtually temple serfs.

We can only guess the early stages of differentiation of this class society, mostly on the basis of much later and more accessible evidence from Greece. It would appear to arise by the progressive modification of the *sharing out* of the produce of the village community, supervised by the priests, who managed to appropriate more and more on behalf of the god, and by the accession into the population of a number of disfranchised men and strangers, who had no right to any share at all.

TRADE AND MERCHANTS

The resulting inequalities were further accentuated and made permanent by *trade*, which itself arose out of ritual exchanges and later became a necessity. At first this was effected by simple barter, then by the use of cattle (*pecunia*) as units, or through valuable goods convenient for exchange because of their ready transportability, such as shells, gold, and silver, and finally by credit. The need for specialized traders arose originally out of the need for foreign goods necessitating journeys or even armed expeditions. These *merchants*, originally city or royal officials, later set up on their own and came to live mainly by trade. At the outset the temple of the king was the chief storehouse and bank on which all economic life was centred. There taxes were collected in kind; from it distribution was made of food and raw materials. Most craftsmen were virtually serfs, receiving raw materials and food from their priestly or noble masters and handing over the finished goods, though even in early times there came to be some independent craftsmen who bought their raw materials and sold their wares. Propertyless men sold their labour for wages. Those in need borrowed; those with superfluity lent at exorbitant interest; those who could not pay were sold as slaves.

LAW AND THE STATE

Laws had to be evolved to prevent these transactions leading to losses to the temple or to bloodshed. These laws are among the earliest written documents. In some of them we find everything regulated down to prices, wages, and doctors' fees. Thus in the Code of Hammurabi we find the fee for setting a bone or curing diseased bowels is five shekels for a man, three shekels for a freeman, and two shekels for a slave – the latter to be paid by the owner.

The force behind the laws could no longer, as in hunting or even village communities, merely be the traditional sense of what was permissible or taboo, or even the clan responsibility for the doings of any of its members, to be settled by a feud or composed by a ceremonial payment. For a city where there was social inequality an apparatus of force was required.

In the cities of Mesopotamia the original assembly of citizens, faced with threats of inner or outer violence, gave way to one-man rule either in the form of the *ensi* or chief temple administrator, or of the *lugal*, great war chief, but also priest of the god. In Egypt the divine priest-king, Pharaoh, was from the first dynasty head of the State. The laws were enforced and the taxes collected by a body of temple servants with police powers. The king also arrogated to himself the right of *punishment* by fine, imprisonment, beating, or death. The power of the State, though vested nominally in an individual, was in fact dependent on the support of the whole of the upper classes of priests and merchants, tempered only by the fear of popular revolt.*

We shall follow in this book the rise and fall – the developments and differentiation – of the *class society* during its 5,000 years of existence. We shall see it in turn as a social form assisting, holding back, or destroying the chances of human advancement. At its very inception, however, there can be no doubt of its generally progressive character. It gave an enormous impetus to the development of techniques and to the beginning of a rational approach to them from which science was to arise.

3.3 The Techniques of Civilization

THE DISCOVERY OF METALS

The organization of river-plain agriculture was the decisive economic factor in the first rise of cities. The major technical advance that accompanied it was the discovery and use of *metals*, particularly that of *copper* and its alloy *bronze*, which has given its name to the whole era of early civilization. However enormous was the subsequent importance of metals to technique and science, they cannot have acquired this importance from the outset. The word for 'metal' comes from a Greek root meaning 'to search', which implies their early scarcity. At first metals were such rarities that they were used only for luxury articles. The agriculture and most of the crafts of the city were carried on by stone techniques. Metal was not even strictly necessary to civilization. None of the great cities of the Mayas or the Aztecs ever knew it except for ornaments; all tools were of stone.

Metals, apart from gold and a little copper, are not found in the raw state, their extraction and preparation imply a long experience and even possibly deliberate experimentation. The original impulse may have come from the interest that primitive man, even in the Old Stone Age, had in all oddly shaped and oddly coloured objects. Bits of metallic ore were bound to attract attention, and in fact have been found in necklaces and other ornaments. It is perhaps more than a coincidence that there was a very considerable trade and use of malachite, the most easily reduced ore of copper, as eye-paint in pre-dynastic Egypt. The use of metals for tools must have been a secondary consideration.

The first of the metals, because it was the most obvious in the native state, was gold.[2.27] But gold nuggets, unlike the hard and brittle stones used for implements, were plastic. They could be beaten out, and a technique of metal-working was developed long before metal could be extracted from the ore. Native copper nuggets, though not so conspicuous or ornamental, could be beaten out into pieces hard enough for tools. This was found to be easier if the metal was first heated or annealed before hammering. This association of metals with fire techniques probably led to the next steps, the reduction or smelting of carbonate copper ores and the melting and casting of the metal produced. Recent research [2.27] seems to show that these steps took place in this order. Both require higher temperatures than can be reached in an ordinary fire, and the evidence points to their association with the production of glazed pots in a kiln with a good draught. A major problem in accounting for the origin of metallurgy is that the localities of native copper or surface oxidized copper ores are usually in hills remote from agricultural centres. It is still an open question whether metallurgy started in the mining areas and the products were rapidly taken up in the cities, or whether both ores and metal were first accumulated in the cities and the technical advances made there. Even if the latter was the case, transport difficulties early in the metal age sent smelters out near the mines.

18. Symbolic figure of a goat about to nibble at a tree. Made of shell, lapis lazuli and
gold, this beautiful work of the goldsmith's art from Ur is of date *c.* 2500 B.C.

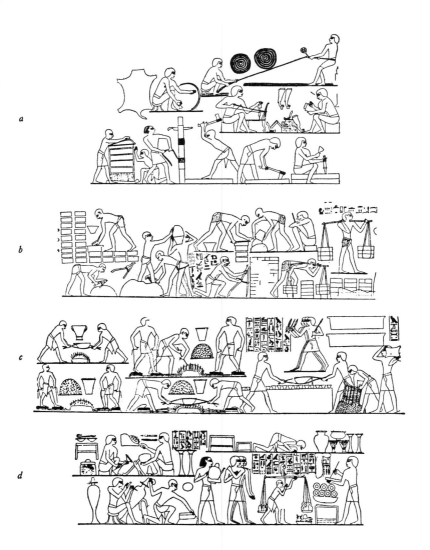

19. Egyptian technology from the tomb of Rekhmiré (*c.* 1470 B.C.). The crafts and work illustrated here show: (*a*) rope making (note swinging weight) and cabinet making (note use of bow drill, chisel and saw); (*b*) brick making and building (note balanced loads); (*c*) bronze casting (note foot-operated bellows and use of tongs); (*d*) finishing vases and weighing precious metals (note similarity of balance to brick carrier).

EFFECTS OF THE USE OF METALS

The production of metal implements and utensils is another technical advance marking a new qualitative change in man's control of his environment. Metal tools are far more valuable and durable than stone tools, and metal weapons are very much more effective than stone ones, both against animals and human enemies. Metal vessels can stand fire without cracking.

On the other hand, metals were for centuries very expensive. Copper ores are sparsely distributed in distant and inaccessible places and tin ores even more so. Both are needed for *bronze*, with its low melting point, which made *casting* feasible. Bronze is far harder than copper, and its use made metal superior to stone for all tools and weapons. Metals and their ores imply distant trade, and with it the inevitably high cost of primitive transport; this must have added very much to their price in the city. Consequently their use was restricted at first to adornment for the temples, utensils for the king's table, tools for the city craftsmen, and then, as war became more common, for weapons.

THE CRAFT OF THE SMITH

The techniques of metal-making and the use of metal tools were of enormous importance to other techniques and in enlarging the craftsman's knowledge of the physical and chemical properties of matter. Sheet and wire were made by hammering and drawing, and casting, welding, soldering, and riveting were developed extraordinarily quickly. These techniques were used to create rich and complex ornaments, vessels, and statues. Because metal-working in bronze, silver, and gold was developed at a relatively late period, unlike pottery and weaving, it was specialized from the start and seems to have been in the hands of close guilds of *smiths*. This was an occupational clan, one early example of what was to become the whole system of minor *castes* in India. The metal-workers must have had a very close guild, as many of their processes remained secret till recent times, or have been lost because no written record was left (p. 596).

The early smiths, apart from those involved in mining and smelting, were mainly concerned with working up metal from ingots or scrap. Most of them must have lived in the cities, but we know from the hoards of scrap and half-made tools they left behind that they must also have travelled around the country like a superior kind of tinker.[2,27]

The value of metal tools and weapons did not lie only in their durability. The fact that a metal tool could take a much thinner section than stone made it cut clean, not merely notch or break. Thus the use of metal tools, particularly the *knife*, *chisel*, and *saw*, transformed the working of wood and made jointed *carpentry* and coursed *masonry* practical on a large scale. The first machines, particularly the wheeled cart and the waterwheel, were only possible thanks to metal. Even in the basic craft of agriculture the ox-drawn hoe or plough became fully effective only when metal replaced stone for its earth-breaking *share*.

TRANSPORT

The mechanical inventions of the first civilization were destined to have immediate as well as long-term effects. The very existence of the early cities depended on the ability to organize the effective transport of materials in bulk. Food from the countryside was needed for thousands of people in the city; trade goods had to be exchanged with other cities, and metals, wood, and even stone had to be fetched from distant forests and mountains. This led to great improvements and radical innovations in the means of transport which were to have far-reaching consequences to civilization, and especially to the growth of science.

THE SHIP

As the early civilizations grew, first round large river valleys and their associated deltas and lakes, they must from the beginning have depended mainly on water transport. Under the stimulus of that need, the primitive dug-out canoes, the bundles of reeds, or the bamboo rafts were built up by almost imperceptible additions, continually tested by practice, into

20. Model of Egyptian boat with single pole mast. This wooden model with rowers was made in Egypt *c.* 2000 B.C.

serviceable *ships*, capable of conveying goods in bulk.[2.49] Indeed, the early political unification of Egypt was made possible and even necessary by the use of the Nile as a waterway. Early boats and ships were propelled by paddles or oars and were to continue to be so for many centuries. However, at some date around the beginning of civilization came another crucial invention – that of the *sail*. This enormously increased the range of shipping; but it is of prime importance as the first application of inanimate *power* to human needs, the prototype of the wind- and watermills of the steam engines and aeroplanes that were to come later.

The rivers and lakes were the training grounds for venturing on the sea, though here the fishermen may have anticipated the traders. Sea travel in turn imposed new problems in ship-building, demanding much firmer construction than was needed for river craft. Further – and this was a point of the utmost importance to later science – it imposed a need for finding the way when out of sight of land. The most primitive method was that of the land-finding bird, as in the legend of Noah's ark.* Land-finding by the stars implies some idea of a map. *Navigation* by sun and stars was second only to the calendar in its demands for practical astronomy.

THE WHEEL

As significant for the future advance of technique and science was the development of land transport which combined two critically important ideas: the use of *animal power* and the *wheel*. Animals had been tamed and bred, at first for food, to satisfy more amply the old hunter's needs. Now they had a new function in doing work, in drawing wheeled *carts* and in taking the place of women in pulling a hoe, thus transforming it into a plough.†

The first use of animals for transport was probably with the pack-saddle. Early man, to judge from the absence of any pictures of him doing so, must have been very chary of riding even donkeys. After the pack-saddle may have come the *travois* – a pack tied to two poles trailing on the ground – still used by some Siberian tribes. This invention does not, however, seem to be the origin of the cart, for in the earliest of these we find the yoke and pole of the plough rather than the shafts of the *travois*. The need to move heavier objects that could not be broken up into loads, like tree-trunks for beams, or stones for great buildings, came only with the rise of cities. For this the first solution was the *sled*, probably only an enlarged version of the light sleigh of the forest hunters. Heavy sleds could be eased downhill, but along the level tree trunks came in handy as rollers.

21. A bullock cart (*eekha*) in India. Leather straps hold the wheel supports – a primitive but effective method of holding a moving roller.

The crucial transition between roller sled and *cart* was probably a city-inspired one, though, once made, the cart spread rapidly to country districts. The real ingenuity lay in securing a solid roller to the body of the cart so that it could turn without coming off. In early Mesopotamian carts and some Indian carts to this day the axle turns with the wheels and is held in place by leather straps. This was the first true *bearing*, though the *door*, with its post and socket, must have run it close. The next stage came in enlarging the ends, first with solid baulks to make *wheels*, and devising a leather and then metal tyre to hold them together. The first development of the wheeled cart seems to have been by the Sumerians, possibly before they came to Mesopotamia. The Egyptians, whose cities were never more than a few miles from the Nile, used boats for most transport; wheeled vehicles were introduced very late.* The light, spoked wheel for war chariots, turning freely, came much later, near the end of the Bronze Age, for it required the extremely accurate joining of the *wheelwright*.

These inventions were to have enormous material and scientific consequences. The cart and the plough between them enabled agriculture to be spread over all open plains and so far beyond the limits of the old civilizations. The two-wheeled ox cart of the Early Bronze Age was the early prototype of the covered wagon that was, 4,000 years later, to open the prairies of the New World. In level country, wherever plough and cart could be used, they added to the effective surplus of agricultural produce, as well as making possible foreign imports in bulk. The lever and inclined plane, already used in the great constructions of temples and pyramids, had laid the foundations of *mechanics*. The use of the wheel, from which were to come water-wheels and pulleys, was to build on these foundations a new edifice of theory that could reach from earth to the wheeling skies. The twelve spokes of the sacred wheel marked off the months of the year, while the wheel itself in motion became the sun-cross or swastika, a symbol first of innocent then of sinister portent. At the same time the increased possibilities and speed of transport by cart and even more by ship, together with the need to know the sources of valuable materials, led to deliberate exploration and to the beginnings of *geography*.

The invention and the subsequent development of all these new techniques furnished an enormously extended field for scientific understanding, just at the time when the organizational needs of the new civilization were bringing into being the intellectual means by which that understanding could be expressed and transmitted.

3.4 The Origin of Quantitative Science

RECKONING, WRITING, AND SCIENCE

The wide scope of operations and the large quantities of materials and services involved in operations of the city temple provoked this qualitative change which marks the beginning of conscious science. In the first place, when they could no longer trust to their memories, the priests were obliged in some way to record the *quantities* of goods received and handed out. This implied the use of *measure*, first as a mere convenience – baskets of grain, jars of beer, pieces of cloth – but then, in order to make them comparable, some standardization was necessary. A set of definite temple or royal measures was adopted and gradually, for the benefit of foreign trade, partly co-ordinated between different cities. Probably later, but still very early, is the measure of *weight* implying the use of

balance with its incalculable consequences for science. The balance must be a city product; in village economy there is nothing that cannot be counted or measured – a shoulder of mutton, a load of wood. It is in the first place required for valuable metal which cannot be measured and where a 'piece' is too indefinite, so that weights are needed. The balance, the only way of comparing weight, bears all the marks of being a *scientific* invention. Its prototype was probably the pole and basket load carrier *balanced* on the shoulder. It needed, however, considerable reduction in scale to be really useful for weighing precious metals.*

NUMBERS AND HIEROGLYPHICS

Even before the standardization of measure it was important to record the *numbers* of objects, whether they were heads of cattle or baskets of grain, that were being collected or handed over. At first this would be done by mere cuts on a stick (plate 7(*e*) p. 83), then by single strokes on a tablet or lump of clay, then by more elaborate designations of large numbers. For the records, where what was in question might have been forgotten, the number symbol was followed by a picture or shorthand symbol of the particular object to indicate what it was that was being counted.

By extension these symbols came to cover actions as well as objects and so to stand for words, either by their meaning alone, as in Chinese, or in part sound–part meaning combinations, as in Mesopotamian cuneiform or the Egyptian hieroglyphics that seem to have been inspired by it.[2.29] The final simplification of the true alphabet, where the symbols stood for sounds alone and not for words, did not occur till the Iron Age. In this way *writing*, that greatest of human manual–intellectual inventions, gradually emerged from accountancy. As Speiser put it, 'Writing was not a deliberate invention but the incidental by-product of a strong sense of private property.'[2.79] First, official statements in the nature of propaganda, praises of kings, hymns to the gods, and last of all science and literature, came to be written down.

22. A low-relief carving of the scribe Hesi-re, *c.* 2750 B.C., from a wooden panel that formed part of a false door in a tomb. The figure holds a palette and other items for writing on papyrus. Now in the Cairo Museum.

MATHEMATICS, ARITHMETIC, AND GEOMETRY

But *mathematics,* or at least *arithmetic,* came even before writing. The manipulation of the signs for objects (as simple symbols) meant that it was possible for the first time to perform the elementary operations of addition and subtraction without counting the real objects in the field. For this it was a matter of matching one collection of objects against another. First came the standard collection, the ten fingers of the two hands, the *digits* of arithmetic, the origin of the *decimal* system. In a pyramid text the soul of an Egyptian pharaoh is challenged by an evil spirit to show that he can count his fingers and triumphantly passes the examination. For more complicated counting, and for adding and subtracting, stones (*calculi*) could be used, which gave us the term for all our *calculations.* Later they were replaced by beads arranged in tens on wires, making the first and still very useful calculating machine, the *abacus.* The introduction of measure made it possible to extend adding and subtraction to quantities. The more complicated operations of multiplication and division come when shareable quantities were involved,

23. The earliest and, in the Far Eàst, the everyday calculator, is the abacus. This example from China shows the principle of operation well. The 'rings' in the upper half each represent five, those in the lower half unity. To express any number, the rings are moved to the middle bar.

particularly quantities connected with public works – the digging of canals, the building of pyramids.

The operation of building itself also contributed, probably even before land survey, to the foundation of *geometry*. Originally, town buildings were simply village huts made of wood or reeds. In cities, with a restricted space and danger of fire, houses of pisé or rammed mud were a great improvement. The next step was to have even greater consequences: the invention of the standard moulded block of dried mud – the brick. The brick may not be an original invention, but a copy, in the only material available in the valley country, of the stone slabs that came naturally to hand for dry walling in the hills. Bricks are difficult to fit together unless they are rectangular, and their use led necessarily to the idea of the *right angle* and the use of the *straight line* – originally the stretched line of the cord-maker or weaver.*

The practice of building in brick, particularly of large religious buildings of pyramid form, gave rise not only to *geometry*, but also to the conceptions of *areas* and *volumes* of figures and solids reckonable in terms of the lengths of their sides. At first only the volume of rectangular blocks could be estimated, but the structural need for tapering or battering a wall led to more complicated shapes like that of the pyramid. The calculation of the volume of a pyramid was the highest flight of Egyptian mathematics and foreshadowed the methods of the integral calculus.[2.62]

Also from building came the practice of the *plan to scale*. Such a plan for a town together with the architect's rule is for instance shown in the statue of Gudea of Lagash in *c.* 2250 B.C.[2.45.265] With these mathematical methods an administrator was able to plan the whole operation of brick or stone building in advance. He could estimate accurately the number of labourers wanted, the amount of materials and food they would need, and the time the job would take. These techniques were readily extendable from the city to the country in the lay-out of fields, the calculation of their areas, and the estimate of their yields for revenue purposes. This is the origin of *mapping* and *surveying*. It was this practical use that later gave rise to the term of *geometry* – land measurement. *Mathematics*, indeed, arose in the first place as an auxiliary method of production made necessary and possible by city life.

ASTRONOMY AND THE CALENDAR

The ability to count and calculate, derived from practical needs of the temple administration, was of immediate use to them in another of their capacities: the making of calendars and the development of *astronomy* which this entailed. Early man must have paid some attention to the sun, moon, and stars, but was apt to be more concerned with the violent performances of the heavens, such as thunderstorms, than with the completely reliable and regular phenomena of day and night. Such a calendar as he needed was provided by the moon, around which had collected much ritual and myth,[2.83] but which at first made little call on mathematics or astronomy.

With the advent of agricultural civilization, the year rather than the month became important. When agricultural operations had to be planned on a large scale it was necessary to know when to start getting ready to do them. Of course Nature often gave fairly good intimations. The first of these, which were afterwards debased into the superstition of *augury*, came from the very practical linking of the birds with the seasons. The cuckoo is significant because he heralds the spring. He may even be thought divine for bringing it. An acute observer of Nature has a fairly good calendar without bothering to count the days at all.

However, in at least one place – the Nile valley – the flood is a regular annual phenomenon for which it is essential to prepare beforehand. The actual length of the year, 365.2422 . . . days, is not easy to find. It requires prolonged and careful observations of the sun and the stars. Such observations were carried out by the priests in Egypt, and already in *c.* 2700 B.C. led to the compilation of a solar calendar that continued in use for thousands of years.

The Sumerians and their successors in Mesopotamia were too attached to the moon to accept such a simple solution. Instead they tackled the far more difficult task of reconciling the lunar and solar calendars. This required recorded observations extending over many generations and the development of accurate computations. It was here that there developed the sexagesimal system – 360 degrees in a circle (near enough to the days in a year); 60 minutes in an hour; 60 'second' minutes in a minute – which we still use for angle and time measurements. These calendrical computations were carried out by means of extensive *mathematical tables*. These tables are elaborations of those used for business accounts. From them has come much of our *algebra* and *arithmetic*, including the all-important place notation that was to return centuries later as the Arabic (Babylonian, Persian, Hindu) numerals we still use.[2.62]

ASTROLOGY

The practice of observation carried out in the temples of all the ancient civilizations, including those of America, extended far beyond the needs of the calendar. The sun, as the regulator of the year, the bringer of the harvest, came to be worshipped as a god. The moon, though ousted from the primacy that it had in the time of the hunters, was not neglected, and observations were extended to the brilliant erratic stars, the planets, that acquired minor divinities of their own.

All this was far more than agriculture or even navigation required, but by then the calendar and the astronomy needed to draw it up had acquired religious significance. The calendar itself was necessary to fix an ever more complicated set of religious holy days, the scrupulous observance of which, as with our own *Sun*day, was considered essential to the preservation of the order of Nature.*

24. The Egyptian cosmos consisted of a flat Earth and a hemispherical sky. It was polytheistic in concept, the goddess Nut represented the heavens, the god Keb (reclining at bottom of illustration) the Earth, while standing in the centre is Shu, who lifted up Nut to the eternal embrace of Keb. Shu separated Nut and Keb each morning, and they returned together at night. Two souls or 'spirit' gods stand each side of Shu. From the Papyrus of Nesitanebtashru (Greenfield Papyrus) described in illustration 10.

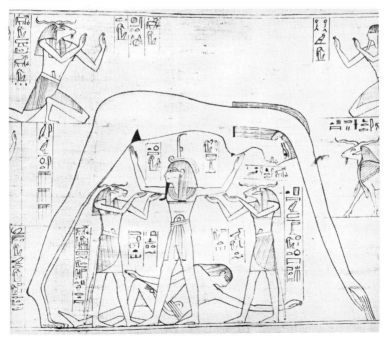

Astronomy was finding other uses. Its study was, from the start, linked with religion. It dealt with the sky-world in which the spirits, particularly those of the sacred kings, lived after death. At first the sky-world was pictured very much like the world below. The Egyptians thought of it as a flat cover, resting on the hills, through which flowed the celestial Nile – the Milky Way. The Babylonians at first pictured it as the inside of a vast four-square tent from which the stars hung like lamps.* It was only after the invention of the wheel that the turning of the sky on its axis round the pole could be accurately imitated. Chinese astronomy seems to have started from this idea of rotation. This is shown by the antiquity of the *pi*, a wheel-like object representing heaven which can actually be used to fix the position of the stars of the Plough. Chinese astronomy retained the dominance of the circumpolar stars rather than the ecliptic for many centuries.[3,8]

The idea of regular rotation of the heavens led to a great emphasis on the movements of the heavenly bodies. It was argued that if these regular recurrences in the heavens affected Nature and brought about the seasons, they must equally affect the condition of man. At first it was only the divine king who was *en rapport* with the skies; but ultimately the privilege became more common, and every individual who could pay might regulate his behaviour by the stars. The seven planets were completely domesticated and still preside over the days of the week. Even their order – Sun, Moon, Mars, Mercury, Jupiter, Venus – was originally astrological. Astrology was always intimately connected with astronomy and, in spite of its essential fallaciousness, it was the major reason why men occupied themselves for millennia with observations of the stars, which, had they not believed in astrology, would have seemed very remote and ineffectual.

MEDICINE

The other occupation that shared with astronomy the distinction of being an upper-class profession was that of medicine. But here, although the prestige was probably as great, the real success, because of the essential complexity of living systems, was bound to be much less. There was in fact practically nothing that a doctor of those times could do except deal with some obvious wounds, dislocations, and fractures, and try to prevent the patient from killing himself, or his relations from killing him by unsuitable treatment or diet. Where the doctors could succeed, however, was in diagnosis. They had in the city enough cases to enable them to compare one with another, and such comparisons, extended by conversation and codified by tradition, are themselves a beginning of science Doctors, long before writing, carried on their traditions orally, first in closed clans which could then be widened by teaching and adoption (p. 186). From the noticing of diseases, and even the recording of them – for we have some extraordinarily interesting examples in early Egyptian papyri[2.13] – arose the sciences of *anatomy* and *physiology*.

Prognosis – knowing how the disease is likely to end – was specially important in early times because the laws, at least of the Babylonians, show that an unsuccessful doctor was likely not only to be prosecuted but even to have his eye put out if by any mistake he destroyed the eye of his patient. It is therefore not surprising that many of the descriptions of cases in an Egyptian papyrus end with the words 'case not to be treated'.

Official medicine codified the plants and mineral substances, knowledge of which had been handed down traditionally from the medicine men and wise women of primitive cultures. Some of these had been chosen for their manifest action as purgatives or emetics; others because in a more obscure way they had been found beneficial in some diseases, as the South American Indians nad found quinine for malaria; but the majority were probably pure magic, based on resemblances such as that of the mandrake to the human body. The city doctors, however, could call on a far larger area for their drugs and could organize their production. It was from this source, rather than agriculture, that arose the science of *botany* and the first botanic or herbal gardens.[1.70]

EARLY CHEMISTRY

Chemistry never rose to the rank of a recognized science in the Bronze Age, or even till near the end of the Iron Age. Nevertheless its basis was being well laid in the multiple observations and practices of the metalworkers, jewellers, and potters. The process of smelting ores, of purifying metals, of colouring them, of adding enamels – all involve complex chemical reactions that had to be learned by many trials, mostly unsuccessful. The good results were embodied in recipes which had to be carefully handed down and scrupulously followed. We do not by any means know yet the full range of the achievements of these early chemists, but what is known is impressive enough.[2.66]

They were acquainted with at least nine of the chemical elements – gold, silver, copper, tin, lead, mercury, and iron,[2.27] as well as sulphur and carbon – and were using and distinguishing the compounds of others like zinc, antimony, and arsenic. They also knew a variety of reagents, dry and liquid, including alkalis like potash and ammonia (as fermented

25. The mandrake (genus: *mandragora*) has a short stem with thick, fleshy forked roots and lance-shaped leaves. The plant has both narcotic and emetic properties and since its roots resemble somewhat the human form, in Roman and medieval times it was believed to possess magical powers. There was a ritual concerned with uprooting it: it must be pulled by a dog from within a magic circle drawn on the ground, and there was a common belief that it shrieked as it came out of the ground. From a wash drawing by H. Killmaurer, sixteenth century.

urine) and alcohol as beer or wine. Their apparatus was limited to pottery and metal vessels; they had no stills and could not cope with spirits or gases.

One powerful impulse was to turn their method of work in the direction of a rational and *quantitative* science; namely the scarcity and value of the materials they dealt with. From the beginning precious metals had to be weighed and accounted for and the proportions used in alloying recorded and adhered to. *Chemical analysis* or assaying, involving the separation of metals already alloyed or mixed in ores, arose naturally out of the necessity of recovering the most precious metals and guarding against adulteration. It was a crucial step in the history of chemistry, and although we cannot precisely date it, we can tell when it arose by the appearance of objects of refined gold instead of the natural gold-silver alloy, electrum. We know from later sources some of the processes used, such as that of antimony for separating silver from gold, and cupellation for separating lead from silver. The astonishing success and persistence of these methods are brought out by the fact that the recipe for a cupel in an ancient Egyptian papyrus – namely bone ash moistened with beer is still the recommended way of making cupels. The astonishing sight of the shining bead of live silver suddenly appearing from the mass of dull, mortified lead oxide made a deep impression. It became the centre of alchemical interest, and inspired spiritual analogies of purification by fire and the resurrection of the body glorious. Indeed, this may have been the origin of cremation (p. 179).

Because we have no works on ancient chemical theory it does not follow that it did not exist. Though it may never have been formally expressed, the ancient chemists show in their products that they were acquainted with the general principles of oxidation and reduction and could introduce or remove non-metals, such as sulphur and chlorine.

As they were mainly concerned with making ornaments they understood particularly well how to produce colours, and since it was the appearance that mattered, they gauged the result by what it looked like.

In trying to make copper look like gold, they produced *brass*; in trying to make the blue turquoise or lapis, they produced a blue *glaze* that was the origin of *glass*. The fact that they were masters of many startling transformations led them to consider that nothing was impossible to their art. This healthy scientific optimism was to degenerate later into the mystical superstition of *alchemy*.

The early chemists never thought of themselves as such, but as metal-workers, goldsmiths, and jewellers. They were highly valuable techni-cians, closely associated with the priesthood and the court, but they were hand-workers at a particularly dirty trade. Their knowledge could not be presumed to be a science on a par with astronomy, mathematics, and medicine. It was an art, but the black art with great magical possibilities (pp. 222, 279).

3.5 The Class Origins of Early Science

It will be seen even from this abbreviated sketch of the scientific achieve-ments of the early civilizations what enormous advances necessarily followed from the foundation of cities. It should also be clear that the scientific, as distinct from the technical advances, were limited to those arising out of the problems of large-scale administration. They were therefore developed by the priests, and also restricted to them, because only the priests had access to the means of recording and calculating. The very term hieroglyphics – priests' writing – brings home that limitation. The association of learning and science with one class in the newly formed class society was to remain its outstanding feature, with a few significant exceptions (pp. 1244 f.), down to our own time. The prestige of mathematics, astronomy, and medicine as noble sciences of the ancient civilizations so impressed the Greeks, and after them the people of our own Middle Ages, that, with the minor addition of music, they remained the pillars of higher education, while baser sciences such as chemistry and biology have still to struggle for cultural recognition. Further, the main programme of science until the eighteenth century, the understand-ing of the motions of the heavens and their connexion with the vicissi-tudes of life on earth, was already established in outline almost from the beginning of ancient civilization.

One significant feature of techniques and culture in the early city States was the extreme rapidity of their development even judged by modern

26. The most famous of all the monumental pyramids are those at Giza, west of Cairo. Constructed around 2600 B.C. as tombs, the largest was originally some 480 feet high, measured 755 feet at the base, and was covered with smooth limestone. The blocks of stone used in their construction were slid into place, it is thought, by lubricating with a liquid gypsum mortar.

standards. For example, it is known that the construction of the pyramids of Giza, with their enormous size, geometrical and astronomical accuracy, and flawless masonry, evolved from that of simple rock-cut tombs in a matter of two or three centuries from *c.* 3000 to *c.* 2700 B.C. Such a speed implies, as does the character of the work itself, the existence of able and practical men, willing to invent and try out new methods over an enormous field of activities. At first it would appear that the innovators were themselves technicians; the legends of such culture heroes as Imhotep, Tubal-Cain and Daedalus show them as craftsmen who both invented and made wonderful new things themselves.

SCRIBES AND WORKERS

But soon after the foundation of the first cities, about the era of the first dynasties of Egypt or the early kingdoms of Mesopotamia, it is apparent that the needs of large-scale organization were already leading to the divorce of the organizers from the actual technical processes themselves.

As they became more numerous and indispensable, they became a caste, markedly separate from the craftsmen and with a great feeling of superiority to them. A very interesting example of this new attitude is shown in a fragment from an Egyptian papyrus of uncertain but early date. It purports to be the instruction of a father to his son whom he is sending up to a 'College for teaching scribes':

> I have considered violent manual labour – give thy heart to letters. I have also contemplated the man who is freed from his manual labour, assuredly there is nothing more valuable than letters. As a man dives into water, even so do thou sink thyself to the bottom of the Literature of Egypt. . . . I have seen the blacksmith, directing his foundrymen, but I have seen the metal-worker at his toil before a blazing furnace. His fingers are like the hide of the crocodile, he stinks more than the eggs of fish. And every carpenter who works or chisels, has he any more rest than the ploughman? His fields are wood, his tools of tillage are copper. Released from his work at night, he works more than his arms [during the day]. At night he lights a lamp . . .
>
> The weaver sitting in a closed-up hut has a lot that is worse than that of a woman. His thighs are drawn up close to his breast, and he cannot breathe freely. If for a single day he fails to produce his full amount of woven stuff, he is beaten like the lily in the pool. Only by bribing the watchmen at the doors with [his] bread-cakes can he obtain for himself the sight of sunlight. . . . I tell thee that the trade of the fisherman is the worst of all trades; truly he does not exist by [his] work on the river. He is mixed up with the crocodiles, and if the papyrus clusters are lacking he must cry out [for help]. If he is not told where the crocodile lurks, fear blinds his eyes. Verily there is no occupation than which better cannot be found except the calling of the scribe which is the best of all.
>
> The man who knows the art of scribe is the superior through that fact alone, and this cannot be said of any other of the occupations I have set before thee. Verily each worker curses his fellow. No man says to the scribe, 'Plough the fields for this person'. . . . One day [spent] in the chamber of instruction is better for thee than eternity outside it; the works thereof [endure like] the mountains. . . . Verily the goddess Rennit is on the way of God. She is by the shoulder of the scribe both on the day of his birth and when, having become a man, he enters the Council Chamber. Verily, there exists no scribe who does not eat the food of the King's House (life, strength and health to him!).[2.14]

It will be seen that already white-collar, or at least white-skirt, occupations are considered to be morally and practically superior, and even worth the intense labour of coping with the fantastically complicated writing and calculating systems of early civilization. The priest-administrators, separated from dealing with material things, tended to elaborate their own symbolic methods and to impute an independent reality to them. In one sense this was valuable, since it gave at least a few select

27. The superior calling of the scribe who had to deal with a complex system of writing and of calculation is evinced by the fact that in Egypt, for instance, it is the ibis-headed god Thoth who is writing with a reed on his palette, and below him, with another palette, is the jackal-headed god Anubis. They are weighing the heart of Nesitanebtàshru who stands naked on the right, her heart in the right-hand pan of the balance and the figure of the goddess of truth in the other. From the Papyrus of Nesitanebtàshru (Greenfield Papyrus) described in illustration 10.

minds the leisure to think, and indeed they were able to create from those symbols the abstract constructions of mathematics. The great achievements of Egyptian and Babylonian reckoners were the foundations on which the later and more abstract mathematics of the Greeks were built. Nevertheless, this preoccupation with symbols permitted the retention of much more primitive ideas, such as the sympathetic magic of the hunting days, and a further enhancement in the power of spirits.

MAGIC AND SCIENCE

Indeed, with the waning of the first impulse of technical advance, magic seemed to become even more important than ever. From being a progressive, if erroneous, explanation of how things work in the world, it became a hindrance to the advance of effective thought. Coming from priests, increasingly divorced from the processes of production, it purported to find solutions to real problems which were seemingly far too easy. By relegating the control of health or success to spirits it prevented the search for useful actions to secure them. It also favoured the use of loose analogies as supposed explanations of natural events in terms of the actions of divine spirits. The world of Nature was seen as merely an enlarged version of the world of man. In fact, every advance of human technique was an invitation to try to understand the rest of the universe in terms of such successful human activity. The major creation myths offer just such explanations. The making of the world is likened to the work of a supreme irrigator separating land from water, and the making of man to the work of a supreme potter moulding him out of clay. Such myths are even more *technomorphic* than *anthropomorphic*.

With due allowance for the enormous difficulties of formulating general scientific theories before the working out of scientific language, we may recognize in many myths the prototypes of scientific theories. In them the forces of Nature are personified, but perhaps their priestly authors took the personification as a mere manner of speaking. Certainly the theories they contained were easily sensed by the Ionian Greeks, and retold without the gods (pp. 173 f.).[2.85; 2.30]

Until science had advanced to the point at which the major part of the environment that mattered to mankind was controllable rationally by direct action – and that achievement is a very recent one – it was, however, very difficult to check the failure of the spirit theory to give man any practical control of Nature. The spirit way seemed no worse than any other, and, by a judicious combination of faith and probability, could even be imagined to work very well. People usually recovered from diseases, crops usually grew, and the sun could be counted on to rise every morning.

However, so long as men held to spirit explanations of natural phenomena, the growth of science was actively inhibited. For not only was any attempt to achieve rational understanding and control deemed from the outset to be useless, but also it might even be harmful, as the spirits would undoubtedly feel annoyed at such attempts to cheat them of their prerogatives. This is only another way of saying that it endangered the livelihood of the priests, who had a vested interest in a spiritual magical theory of the universe, especially when the early temple establishments decayed, and the priests became increasingly dependent on the offerings of the faithful.

The danger to the aristocracy of the gods of attempting to control the forces of Nature is the fundamental meaning of the Prometheus myth.[2.82]

Fire from the first belonged to the heavens; man had no right to take it for himself. What the priests wanted was *piety* – unremitting practice of the rituals of propitiation, the careful observance of all taboos, and resignation to the will of the gods. As long as these views were supported by authority – and they have not yet disappeared from society – it was even impious to inquire too closely into the method of working of the universe. Such inquiries were bound to be resented by the heavenly powers, and their resentment would be vented not only on the inquirer but on the whole of society. The forces of religion were, from the beginning, closely identified with the maintenance of class rule. When, some centuries after the first founding of cities, the ruling classes ceased to favour material and technical progress, religion was bound to restrict intellectual advance.

3.6 Successes and Failures of the First Civilizations

Taken as a whole, however, the early civilizations did succeed in making and sustaining an enormous advance in techniques and, ideas. The high level of their technical achievements is shown by the fact, so common that we hardly pay attention to it, that for the greater part of our lives we are surrounded with and use equipment evolved at that time and scarcely altered in the intervening 5,000 years. Our chairs and tables have not

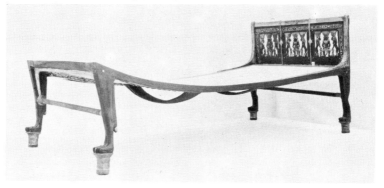

28. An ebony bed from Tutankhamen's tomb, with legs inlaid with ivory and a headboard with carving depicting the domestic god Bes. The stretchers underneath the bed are curved to allow for the sagging of the woven string mattress. Constructed *c.* 1350 B.C., it still shows no noticeable warping or deterioration.

changed since the first Egyptian carpenters solved the difficult problems of wood joinery. Armchairs with wicker seats and claw feet are known from *c*. 2500 B.C. We still live in rooms with walls and ceilings of stone, brick, and plaster; we eat from the same kind of dishes; we wear clothes made of the same kinds of cloth.

Even our social institutions have not changed to an extraordinary extent – far less than the change between the institutions of primitive communities and of the first cities. We have merchants, magistrates, and soldiers just as they had; and the political troubles of our time were not unknown to them. In other words, most of us are still living in the class society that originated with the first cities.

TECHNICAL STAGNATION

The great burst of technical innovation that came with the beginning of city life in the great river valleys of Mesopotamia, Egypt, India, and China did not last more than a few centuries, roughly from 3200 to 2700 B.C. It was followed by a relatively far longer period of cultural and political stagnation. Particular cities rose and fell; one dynasty of priest kings superseded another. There were irruptions of barbarians and even barbarian dynasties, but there was no essential change in the pattern of production. It remained based on irrigation agriculture, supplemented by trade with outer regions. All the wealth that was accumulated and consumed in the cities came from the surplus of city-directed agriculture. Because the surplus was relatively small, only comparatively few of the people could be supported on it, and these tended to form an exclusive class. The successors of the original administrators who worked to improve agricultural techniques became increasingly divorced from the process of production. Their only interest was to secure as much of the product as possible for themselves. From generators of wealth, they became exploiters. They demanded ever more and more for their private enjoyment and for the building and service of increasingly magnificent temples and tombs. This meant the impoverishment and virtual enslavement of peasants and urban craftsmen, and led to conflicts which weakened the city States and ultimately put a stop to their intellectual and technical progress.

29. Statue of Gudea of the Sumerian city of Lagash, *c*. 2250 B.C. This postdates by a century and a half the social revolution mentioned in the text (page 136) but is significant not least for the plan of the city carved on the figure's lap, the stylus (seen in relief) on the left of the plan and, below the map (on the bevelled edge), a ruler with different scales.

Of one of these events we possess fairly full details. In the Sumerian city of Lagash, in its time – 2400 B.C. – the chief city of southern Mesopotamia, there occurred what may justly be called a social revolution. A certain Urukagina seems to have seized power from the rulers of another dynasty and set under way a whole series of social reforms designed to limit the oppression of the bureaucracy, the priesthood, and the rich. Records of these have come down to us in which the contrasts between the old and the new deal are emphasized. Graft and corruption were put down and those convicted of them were dismissed, together with a general cutting down of an army of revenue officials and inspectors. The priests were deprived of many of their privileges and the fees they charged for burials, weddings, divorces, and divination cut to a third or less.

The reforms, however, did not last. The new deal did not destroy but only curbed the ruling class, and its members took the first opportunity to ally themselves to the ruler of the rival city of Umma and launch a war in which Lagash was pillaged and destroyed. One of the loyal priests sadly records on a tablet: 'Of sin on the part of Urukagina king of Girsu, there is none. For Lugal Zaggisi patesi of Umma may his goddess Nidaba bear their sin upon her head.'[2.45.176] The success of the conqueror was short lived. He was defeated in his turn by Sargon, the first king of Akkad, the founder of the first world empire, who was, or claimed to be, like Moses, a foundling picked up by a gardener.

WAR

The ending of this story brings out another potent source of unbalance of early city economy – the organized violence of war. The apparent limit to the exploitation of the local agricultural population could be overstepped by an extension of the area of the city. Up to a point this could proceed peacefully, but if several cities were following the same policy in a restricted area it led to conflicts and to the evolution of a new institution, that of war. War, in its full sense, is indeed a product of civilization. The fighting which recurred between tribes in the hunting or even in the pastoral stage was more in the nature of football matches than of sustained campaigns. Although cruel in detail, it could have very little general effect on a culture where, in any case, it was impossible to concentrate large bodies of warriors or maintain them in the field for more than a few days at a time. Once cities existed, however, that situation completely changed; armies could be heavily equipped and be fed from the surplus food stocks. The upper classes who controlled city governments had strong economic incentives for war. Their wealth depended directly on the area they could exploit, and cultivated land could be taken from another city together with the peasants who tilled it. Beyond that lay the possibility of seizing material, animal, and human booty.

War made the recruiting and leading of armies a vital necessity, and this changed the character of government and the State. The principal function of the head of the State changed from that of a director of agriculture and public works to that of a war leader – from priest to king. Another effect was once more to depress the position of women. In the first phase of civilization women had maintained the great importance they had gained in the village cultures. As war became more important their administrative functions were taken over by men, though they never sank to the position of domestic slaves that was to come with the Iron Age.

WARFARE AND TECHNIQUE: THE ENGINEER

As warfare became more the rule than the exception, and the city began to be distinguished from the village by its defensive *wall* and its fortified *citadel*, the direction of technique began to be more and more influenced by the needs of the armies. Even the newly emerging science was bent in the same direction. Technical progress in weapon construction went on even at a time when other technical progress had almost stopped. We have only to think of the prestige attached in legend to such figures as Vulcan or Wayland the Smith to realize the importance of the armourer to the warrior. Even more important in the long run was the invention of military machines, such as catapults and moving towers, which demand an

30. Military machines are no recent innovation, and the wheeled battering ram with armed men in an upper storey was used by the Assyrians (*c.* seventh century B.C.) and, it seems likely, earlier still. A sketch from a low-relief carving found at Nimrud (Calah) Iraq. From Joseph Bonomi, *Nineveh and its Palaces*, London, 1869.

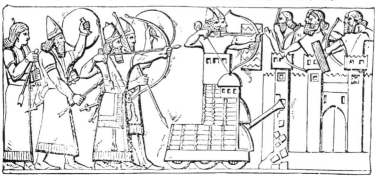

appreciation of the principles of mechanics. The need to make and service such machines, to build earthworks and drive mines, gave rise to the profession of the *engineer*, first and foremost a military profession, though originally drawing his skill from civil sources.

Other and more remote aspects of war also stimulated science. The problems of the supply of armies, including the making of roads and canals,[2.94] were among the most important; so was the design of fortifications, one of the earliest examples of plans to scale. Plato considered that the only practical use for geometry was the drawing up of ranks and files in an army. Without war, or the social system that gave rise to it, the arts of peace could have advanced far more rapidly. But it may at least be said for the association of science and war that war kept science alive at a time when other aspects of culture were decaying.

TRADE AND EMPIRE

Partly by war, partly by a system of alliances based on trade, the originally independent city States tended to become merged into larger units either under the stable and overwhelming preponderance of one city, such as Memphis in Egypt, important not so much in itself as the sacred city of the god-king, or under a shifting sequence of predominance among cities, such as the successive *empires* of Ur, Larsa, Isin, and Babylon in Mesopotamia. In Egypt the concentration of power in the hands of the god-king Pharaoh and his priestly administration (*Per-aah* means big house =Whitehall) was such that enormous and economically unremunerative works like the pyramids could be undertaken. In Mesopotamia the cities were more equal and, though in the aggregate the wasteful expenditure of the upper class may have been as great as in Egypt, it was never so concentrated. In India before the Aryan invasion large city States with citadels, temples, and baths, similar to those in Mesopotamia, existed, but, lacking an understanding of their script, we know too little about them to assess their social structure.[2.67] In early China the prestige of the emperor, the son of heaven, seems to have corresponded more with that of Pharaoh, though for a very large part of its history China has been divided into a number of warring States.

EMPIRE AND THE SUPREME GOD

One result of the growth of empires was the precedence it gave to the god of the ruling town over those of the conquered or federated towns. Amon, originally the ram totem of the nome or parish of Thebes, became, with the rise of the Theban empire, joined to the hawk totem sun-god Ra, as Amon-Ra, lord of the gods. The local god, Marduk, became equally important in Babylon. The power of the god grew and waned with that of the empire, but it left behind the idea of one supreme god as ruler of all the world. Akhenaton tried to realize this idea officially in Egypt with his worship of the sun disc, but failed. It was left for the then obscure tribes of the Jews to succeed and to found modern monotheism.

31. King Akhenaton (Amenhoten IV) the heretic king of Egypt whose reign began *c.* 1360 B.C., his consort Nefertiti and their two daughters offering gifts to the Sun whose outstretched beams symbolize the gift of life. The low-relief was originally in the Temple of the Sun at Thebes, destroyed after Akhenaton's death; this is a photograph of what seems to have been the sculptor's model and was excavated from the tomb of the Princesses at Tell-el-Amarna. Now in the Cairo Museum.

3.7 The Spread of Civilization

While civilization stagnated at the centre its influence was spreading wider and wider. The existence of empires accentuated a problem that must have arisen with the very beginning of river-valley civilization – the relation of the city States to their less advanced neighbours in the open country and the hills. Civilization had provided better techniques, such as those of the plough, the wheel, and the metal sickle, applicable to farming in lands other than those where it originated. It therefore tended to spread in a variety of ways. One of these was simple trekking. The villagers, when the city territory failed to absorb the growing population, moved off with their herds and carts into wilder, less hospitable but roomier country, and thus village communities spread all over the arable lands of Europe, Asia, and Africa and possibly also to America. In this spread many of the more complex products of civilization were necessarily lost or simplified, so that it becomes difficult to distinguish between civilized emigrants who had gone native and people of simpler cultures that had acquired, by transmission from neighbour to neighbour, some of the techniques of civilization.

Another way of spreading civilization was through the trader, and particularly the miner, those more adventurous spirits of the cities who went out into the wild borderland not to settle but to collect local produce of value, especially precious stones, ores and gold. Because the traders had to give city products in exchange they served to spread the needs and, to a lesser extent, the productive methods of civilization. They also inevitably came into conflict with the local population, and invoked the help of their home governments to protect them. This led to a third way in which civilization was spread, the way of political and military interference that we still associate with *imperialism*. The records of the rulers of ancient Egypt and Mesopotamia are full of accounts of punitive or raiding expeditions to the gold mountains, the ivory country, or the pearl islands. Nor was interference limited to military action; as much could be achieved by discovering and making use of mutual antagonisms between foreign tribes or between rival factions inside them. The profession of *diplomacy* long antedates classical civilization.

THE FIRST BARBARIANS

The expeditions led sometimes to actual extensions of settlements under the control of the parent city, as for example, the Babylonian mining settlements at Dûr-gurgurri, though this form of *colony* is much more characteristic of the later Iron Age. The main result was to generate increasingly effective opposition to the city empires. In time the institutions of the peoples in an area hundreds of miles round the centres of civilization had been changed by their intercourse with it. This was the area of the *barbarian* fringe. The barbarians were able to pick up items of the material culture of the cities, especially those which could be easily transported and would involve the least change in their own habits. These were primarily weapons, which, though expensive, would bring in a far greater return than their cost if used in raids on the wealthier centres.

The tribal institutions of the barbarians were also transformed by introducing private property, emphasizing the role of the warrior, and increasing the authority of the chiefs. The effects were greatest in the cultures of pastoral peoples, who were highly mobile, unassimilable by civilization, yet dependent on it for many necessities such as tools and weapons, as well as for ornaments, which they had not the skill to make.[2.2] The relations between the barbarians and the city States were variable and complex. Strong empires played one barbarian tribe against the other, raided and enslaved them. Weak empires were undermined by the importation of barbarian slaves and soldiers.[1.9] In the end they were often completely overthrown and ruled by barbarian dynasties which usually soon acquired the culture of the cities.

SLAVERY

One result of the relations between the city States and the barbarians was the steadily increasing importance of slavery. The institution of slavery, the ill effects of which dog the world to this day, goes back to the beginning of the river cultures. In the days of hunting or early agriculture there was little surplus. A working man did little more than earn his keep. Prisoners taken in inter-tribal feuds, if they escaped being sacrificed, were usually adopted; there was no point in enslaving them.

In civilized countries, on the other hand, an agricultural labourer could produce far more than it cost to keep him. That made the taking and using of slaves an attractive proposition. Slave-raiding from other cities, or more easily and profitably from barbarians, soon became an accepted practice.

The full development of slave-based agriculture was not to come till the Iron Age, but from the beginning of the Bronze Age it had started to exercise its ill effects on civilization. Bound prisoners, intended for slavery, are represented in the oldest Sumerian carvings of about 3000 B.C.[2.45] The existence of propertyless and also rightless slaves was bound to have a depressing effect on the status of free workers. By its association with

32. Low-relief carving of prisoners submitting as slaves. The campaign is that of Assur Bani-pal against the King of Susa, showing the submission of the Susians, captives and damsels playing timbrals, *c.* 668 B.C. Now in the British Museum.

that of slaves their work became base and menial. There was little incentive or opportunity for the free workers and none at all for the slaves to improve techniques, and the upper class scorned them. As a result the scientific approach which had been so successful in the upper-class sciences of mathematics, astronomy, and medicine was cut off from the problems and the information that were to be found in the trades, and for long did not spread to the black art of chemistry or the lowly practices of agriculture.

The political ill effects of slavery were often more immediately disastrous. The more a city depended on slaves the less it was able to look to its own defence, and the more likely were the barbarians to become acquainted, as escaped slaves or later as mercenary soldiers, with the war technique of the cities themselves and to be able to use that knowledge to overthrow them.

For several centuries before their fall, that is, roughly after 1600 B.C., the ancient civilizations of the West, though not those of China, seemed to have lost any capacity for progressive change and were becoming increasingly decadent. Although the framework of civilized life was maintained, art and literature became conventional and religion tended to become buried in an increasingly complex mass of ritual that could appropriately be called superstition. Though much had been lost or forgotten, some science, such as astronomical observation, was kept up and even developed; some degenerated into superstition, such as the careful examination of the livers of sacrificed animals used for foretelling the future. This is only one example of the use of systematic study of obscure phenomena for telling fortunes, cheiromancy (or palmistry), oneiromancy (or dream interpretation), many of which are still with us, either in their original form or in the games of chance and skill like dice, cards, and chess derived from them. In so far as they sharpened the acuteness of observation and the methods of codifying results they have some part in experimental science. One key discovery, that of the compass, was probably made by a Chinese geomancer (p. 317).

3.8 The Legacy of Early Civilization

Nevertheless what remained to be handed on to its successors must have been an impressive and valuable stock of knowledge – far more than the spade of the archaeologist is ever likely to reveal to us. At the same time the archaeologist is certain to know much that would not be known to people living a few hundred years after any event. Although the sources of knowledge may have been forgotten, many of its usable parts are likely to have been assimilated in an unacknowledged form. As the knowledge and practice were alive they could be learned by word of mouth and by the example of the practitioners. Only a certain amount of knowledge was assimilable by new cultures with different social and economic structures. The enormous accumulation of history, poetry, and literature of those times was largely lost with the knowledge of the hieroglyphic and cuneiform scripts in which they were written. The little that has survived in the Bible shows the level to which they reached. Much priestly science must have gone too. Techniques fared better; both the equipment of civilized life and the tools with which it was made largely survived and are in use today (pp. 133 f.).

The science and techniques of the Iron Age and even of the Greeks are largely derived from those of the ancient world, for the most part without acknowledgement. Indeed, in the case of techniques which are embodied in material and durable objects we can be sure that this has occurred. Many ideas or discoveries have been attributed to a Greek philosopher for no better reason than that he was the first known to us to have expressed them or been credited with them. Further research often reveals an earlier origin in Egypt or Mesopotamia, and we have no reason to believe that the present verdicts of archaeology are final.

The heirs of the old civilization, the men of the Iron Age, had themselves no doubt about the greatness and magnificence of the empires that they had helped to destroy. Echoes of the life of those times are to be found in the *Iliad* and the *Odyssey*, themselves tales of city sacking and piracy. The poets contrasted their own hard lives and mean culture with the power, the luxury, the beauty, and most of all the peacefulness, of the old cities. They revered the wisdom of the Ancients and looked wistfully back at the age of gold.

Chapter 4

The Iron Age: Classical Culture

The period covered by this chapter is one of crucial importance in the history of mankind and especially in the history of science. From the middle of the second millennium B.C. a number of causes – technical, economic, and political – brought about the transformation of the limited civilization of a few river-basins into one which embraced the major cultivable areas of Asia, northern Africa, and Europe. The civilization of the Iron Age, wherever it was developed, was less orderly and peaceful than that it replaced, but it was also more flexible and rational. The Iron Age did not provide such enormous technical advances as marked the outset of the Bronze Age, but such advances as it did achieve, based on a cheap and abundant metal, were more widespread not only geographically but also among the social classes.

In this chapter we shall deal primarily with the Iron Age in the Mediterranean area – the classical civilization of the Greeks and Romans. This is partly because it is so much better known than that of contemporary cultures of India or China. A more cogent reason, and one that relates particularly to the purpose of this book, is that it was that Mediterranean region which gave birth to the first abstract and rational science from which the universal science of our own time is directly derived. As we shall see in subsequent chapters, the civilizations of India and China had great contributions to make to the common culture, particularly in mathematics, physics, and chemistry, and their applications, like the compass, gunpowder, and printing. However, the contributions entered into the main tradition of science and technology only after its outlines had been fixed in its Hellenistic form.

4.1 The Origins of Iron Age Cultures

The barbarians who overran the bronze age cultures of the ancient east had been unable to form stable states in their own homelands, for the most part covered with forest or dry steppe, as long as they lacked the means of establishing some settled form of agriculture. In the latter half of the second millennium B.C. these conditions were achieved, thanks to a combination of material and social factors that we are only now beginning to understand. Of these one was the penetration and transformation

33. The olive harvest was an important aspect of Greek agriculture, for it brought a fresh supply of oil. Greek amphora with black figures, sixth century B.C. Now in the British Museum.

of barbarian clan societies through the influence of the class economies of the cities with their emphasis on private property, chieftainship, and weapon production.

THE IMPACT OF THE DISCOVERY OF IRON

These tendencies were powerfully, perhaps decisively, aided by the discovery and use of a new metal, *iron*. Where and how iron was first made in quantity is still a mystery. The first iron used was the native iron from meteorites treated by heating and hammering like copper, but this was too rare to be anything but a precious metal. The first iron to be smelted from its ores was probably a by-product in gold-making[2.27] and must have been even rarer. Iron in usable quantities seems to have first been smelted from the ore somewhere south of the Caucasus by the legendary tribe of the Chalybes, in the fifteenth century B.C., but it did not appear elsewhere in sufficient quantities for its use to be economically and technically decisive until about the twelfth century B.C. The wide distribution of iron and the ease of iron-working ended the monopoly of civilization of the old river empires of Egypt and Babylonia. Two other developments hastened the process – the appearance of mounted *horsemen* from the steppe lands where the wild horse, far more powerful than the ass, had been tamed, and rapid improvements in the performance and building of *ships*, itself a by-product of iron technology.

THE METALLURGY OF IRON

The iron used in antiquity, indeed up to the fourteenth century A.D. in Europe, was made by a process of low-temperature reduction by charcoal in a small, hand-blown clay furnace. The resulting *bloom* of spongy, unmelted pure iron was beaten out into bars of relatively soft *wrought iron* from which more complicated forms could be made by *forging* and *welding*. The first elaboration of the techniques of iron-making and working must have been the fruit of long and difficult experience. This technique was totally different from that for copper and this probably explains why iron metallurgy came so late. Once established, however, it required nothing but the simplest equipment and could be quickly taught or picked up. Wherever there is wood and ironstone – that is, almost everywhere – iron can be made: once you know how.

Iron had one serious disadvantage as a metal in early times: it could not be melted for lack of sufficient blast to the furnace, and casting was therefore reserved for bronze, except in China, where cast iron was made as early as the second century B.C.[2.18] Iron did not displace bronze; it merely supplemented it for common purposes. More bronze was made and worked in the Iron Age than ever in the Bronze Age itself. The iron made by the bloomery and forging process was a wrought iron or very mild steel; it was tough but relatively soft. Much harder true steels were known – chalybs from the Chalybes, ferrum acerrum, sharp iron, acier – but their method of manufacture was kept a deep secret among the tribes of smiths. The world of science was not to know it until the work of Reaumur in 1720 (p. 596). The secret was essentially one of getting more carbon to combine with the iron and then hardening by tempering and quenching. The best steels were those made by the Chinese – seric iron – and by the Indians, whose *wootz* steel [2.27] was exported to make the famous damascened blades. Good steel was so rare and highly prized that the swords made out of it were deemed to be magic, like Arthur's Excalibur or Siegfried's Balmung of later times. Tempered steel was far rarer than bronze and except for use in weapons was to play no important part in technique until the eighteenth century.

The introduction of iron coincided with a period of folk wandering. More or less barbarized tribes came down from eastern Europe or the Caspian into the eastern Mediterranean area from the seventeenth century B.C. onwards. Similar movements of Hittites, Scythians, Persians, and Aryan Indians were occurring in Asia. The great mobility of the horsemen and the sea peoples and their abundance of new weapons made it difficult for the old empires to put up an effective military resistance. We may suspect that military failure was an index of lack of support from the people of the older civilizations, who were more likely to sympathize with the invaders than with their own inefficient and rapacious rulers. Further, the iron age peoples, once they settled down, showed themselves capable of building prosperous agricultural or trading communities on hitherto fruitless land. The result was to reduce the political and economic pre-eminence of the early river-valley civilizations to such an extent that they no longer figured as the main centres of human cultural developments, although many of their cultural, material, and spiritual achievements were transmitted and even some of their records were preserved.

Instead, the effective foci of advance moved to the periphery of the ancient civilizations, the settlements of the nearer barbarians, who had managed to overrun the older centres of civilization but who developed their culture largely outside them. The Aryan Indians, the Persians, the Greeks, and later the Macedonians and Romans fell heirs to the old civilizations of Egypt and Babylonia. The position of China was exceptional; surrounded largely by steppe, desert, and mountainous areas, there was little possibility of the building up of agricultural barbarian

States beyond its boundaries. The nomadic barbarians that repeatedly moved in were all absorbed by ancient Chinese culture. That basically bronze age culture, though profoundly changed by iron age techniques, has retained its continuity right down to our own time.

AXE AND PLOUGH

The destruction and wars of the early Iron Age were, however, not without their compensations. The substitution of new cultures for old meant certain losses of continuity, but it also meant the sweeping away of much accumulated cultural rubbish and the possibility of building much more effective structures on the old foundations. If the mounted warriors and the shiploads of pirates stand as symbols for the destructiveness of that period, the woodmen with their axes and the peasants with their iron-shod ploughs amply made up for the destruction. The earlier use of metal was essentially for the luxury products of city life and for arming a small *élite* of high-born warriors. Bronze was always too expensive for common folk, who still had to rely for the most part on stone implements, the form of which had scarcely altered from neolithic times. Iron, however, though originally, and for many centuries, inferior to bronze, was widely distributed and could easily be produced and worked locally by village smiths.[2.27] The effect of the abundance of iron was to open whole new continents to agriculture; forests could be cut down, swamps could be drained, and the resulting fields could be ploughed. Europe, from being literally a backwoods, became a new 'golden west' – in the sense of its wheatlands rather than of its gold, which was largely exhausted at the end of the Bronze Age. The resultant increase in population rapidly altered the balance of power between the dry farming of the West countries and the old river-irrigated cultivations of the East.

SHIPS AND TRADE

Another feature of the disturbed times of the Iron Age that was to be of incalculable importance to human thought, and particularly to science, was the use of the sea-ways in spreading culture much more rapidly than the old overland routes could possibly do. What was more important was that transport by sea was many times cheaper than by land. With the greater facilities for ship-building provided by iron tools there were better and larger ships and more of them. In the Mediterranean the initiative in shipbuilding had been taken by the Cretans in the Bronze Age. The breaking up of their sea empire, first by the land-based half-Greek Mycenaeans and later by the more barbarous Achaeans from the Balkans and by kindred tribes in Asia Minor, was the signal for a great period of piracy

34. Greek ships: on the left a merchant ship and on the right a war galley. From a black figured Kylix *c.* 520 B.C. Now in the British Museum.

and sacking of cities. The immortal story of Troy records one of these expeditions. Naturally piracy made trade difficult but it also made it profitable, and former pirates, attracted by this or deterred by more effective local defences, gradually turned to trade, exploration, and colonization.

In the Iron Age trade ceased to be a matter concerned only with a round dozen great cities, like Thebes or Babylon, and became more and more divided among the hundreds of new cities that the early iron age peoples, such as the Phoenicians and the Greeks, were founding all over the shores of the Mediterranean and the Black Seas. Only places near the sea could get the full advantage of iron age culture. In countries far removed from it the Iron Age certainly brought greater possibilities for agriculture and warfare, but, where there was no way of moving bulk products over long distances, such countries could not progress economically even as far as the bronze age civilizations with their river transport. They were consequently unlikely to produce anything radically new. The Assyrians, a typical land-based early iron age people, were distinguished mainly for their military ruthlessness. They preserved for some centuries the old Babylonian culture, including the continuation of astronomical observations invaluable for the science of the future,

35. An elegant limestone pillar erected by Aśoka in 243 B.C. at Lauriyā-Nadangarh in Nepal. The lion faces east towards the rising sun and the whole is almost 40 feet high. Aśoka built a highway extending from the river Indus through the Punjab to the Ganges, a distance of some 850 miles. Along this Royal Road were resting places where medicinal plants were cultivated.

36. A hoplite – an infantryman of the Greek city states. Citizens unable to afford horses, yet of sufficient means to buy full personal armour, were required to serve as hoplites. The armour consisted of helmet with cheek pieces, breastplate, shin protectors (greaves) and a strong shield. Socrates served as a hoplite, although he was later reduced to poverty. Bronze statue from Bodona.

but added little to it themselves. This advantage of the sea-way could not be fully offset by roads such as those made first by the Persians and later by the Romans. These were of administrative and military rather than economic value. Land transport in bulk could not begin to be economical until the development of efficient horse harness in the Middle Ages (p. 312). Even then it was not practicable over large distances till the making of good roads in the eighteenth century. It was the ease of water transport that gave first the Mediterranean area and later all Europe, with its indented coastline, an advantage over Africa and Asia.

China, with its network of rivers, canals, and lakes, had some of the same advantages but, as it retained, even through periods of warring States, bureaucratic governments of a modified bronze age type, it missed many of the economic and political developments of the Iron Age.

4.2 Iron Age Cities

POLITICS

In its early stages the Iron Age meant a return to a smaller scale of economic unit. Early iron age cities rarely had populations of more than a few thousands, as against the hundreds of thousands of bronze age cities. By the fifth century B.C., with the spread of slavery, much larger cities were possible. Athens had a maximum population of 320,000, of whom only 172,000 were citizens, while Rome at the height of its power had about a million. The first cities were formed by the agglomeration of a dozen or so villages.[2.83] This, however, did not mean a return to neolithic conditions, but, for the population at large, to one with standards as high, or higher, than that of the Bronze Age. The iron age city had inherited all it could use of the arts of the bronze age cities, that is, all but the organization of large-scale works. Early iron age cities, with their restricted areas, rarely went farther than fortifications, harbours, and occasional aqueducts. Also it had, in addition, the use of a metal that enormously improved agriculture and manufacture and it did not need to be self-sufficing: it could rely on trade for necessities as well as luxuries. This was possible only because improvements in the methods of production enabled goods to be made for a market. The Iron Age is the first in which *commodity production* becomes a normal and indeed an essential part of economic activity.* Another social economic feature of the Iron Age was the use of slaves not merely, as of old, for service but also as a means of producing for the market. This was mainly in agriculture and mines, but it spread also into manufacture. Slavery, as we shall see (pp. 224 ff.), grew steadily in importance till it became the predominant form of labour. This was itself a substantial factor in causing the breakdown of the whole culture, with the consequent turning of slaves and poor freemen alike into a common people of serfs (p. 253).

The iron age city became, almost from its inception, a well-placed centre for manufacture and trade, able to get from abroad its raw materials and even its labour force, as slaves, in return for the sale of its products.

Against these advantages was set the much-increased danger of war. The new culture had been born in war – the sacking of cities in a state of permanent rivalry. It was difficult to outgrow these habits; defence became a priority; cities were built most inconveniently on hill-tops, like the old high city – Acropolis – of Athens; or on islands, like Tyre; and automatically all citizens had to be soldiers. Nevertheless the small iron age city was both simpler and freer than the old river-valley city. It also gave much greater scope to its citizens, who were forced to organize themselves to look after common interests rather than take their place in a preordained hierarchy. In this way the iron age city gave rise to *politics* and created out of political struggles between the classes in the cities the successive forms of *oligarchy*, *tyranny*, and *democracy*.

MONEY AND DEBT

One great social invention that provided both for the expansion and the internal instability of iron age civilization was that of metallic *money* – first as stamped bullion in Lydia, and then, after the seventh century B.C., as coin. Metal by weight had been used as currency in the old empires, but its use was exceptional, and barter and payment in kind remained the rule. Money, which soon became the measure of every other value, turned all established social relations into those of buying and selling. Precisely because of its general and anonymous character, money, by bringing rights without obligations, enabled power to be concentrated in the hands of the rich. At the same time, by superseding the old tribal distribution of real wealth, it took all protection away from the poor. For

37. A black granite stele set up under the young Alexander. It records the deed of grants to the temple of Buto in Delta. The hieroglyphics represent objects or syllables of objects, and should be contrasted with the cuneiform script on the diorite stele that records the laws of Hammurabi (see plate 17). From these forms of written or carved information there arose the alphabets that were to transform the communication of knowledge. Stele now in the Cairo Museum.

them the existence of money was negative; they lived in a state of chronic *debt*. True, the oppression of the poor is as old as civilization itself (p. 130). Nevertheless there were real differences between the forms it took in the old civilizations and those in the Iron Age. In the earlier case it was gradual and partial. The economy had arisen directly from a tribal society and tradition was a bar to arbitrary action. The cultivator had many duties, but he also had rights. If he belonged to the land, the land also belonged to him. His payments were in kind and commercial transactions and debts were largely limited to the city population. In the Iron Age there was an abrupt transition from a clan to a money economy. Immemorial customs were destroyed in a few generations and the rule of money could disregard all rights.

On the other hand, the cultivator was potentially far more independent. If he found the situation intolerable he could join a band to form a new colony. If enough people found it intolerable they could and did revolt. With the common use of iron and the training of all citizens in arms these revolts were often successful, and the fear of them kept oligarchs and tyrants in check.

Nevertheless from the beginning of the Iron Age the oppression of money power and the repeated, but inevitably temporary, successes in breaking away from it by reform or revolution becomes the general background theme of city history. Towards the end of the classical age, under the Hellenistic and Roman Empires, money power seemed to triumph absolutely; but its very triumph led to a state of such widespread misery and hopelessness that the whole system broke down and returned to a simpler feudal economy in which money at first played only a small part.

THE ALPHABET AND LITERATURE

An iron age development of importance for the origin of science was the vulgarization of the elaborate systems of writing – hieroglyphics and cuneiform – of the ancient empires into the common Phoenician *alphabet*, which made *literacy* as cheap and democratic as iron.[2.89] The alphabet arose in relation to trade between people who had different languages but had to deal with the same things. As its symbolism was based on sound it could be applied to all tongues and at the same time it opened the world of intelligent communication to a far wider circle than that of the priests and officials of the old days. Writing ceased to be confined to official or business documents and a *literature* of poetry, history, and philosophy began to appear. Naturally poetry and prose narratives themselves, in the form of *epics* and *sagas*, must have long preceded alphabetic or even hieroglyphic writing, being handed down by bards or professional story-tellers. It cannot be claimed that an alphabet is an essential to the production of literature, as the example of China shows. Nevertheless, the Chinese achievement was only made possible by the creation of a bureaucratic feudal class who monopolized learning and also largely sterilized it.

4.3 Phoenicians and Hebrews

The first peoples to profit by the new conditions of iron age civilization were the Phoenicians of the Syrian coast. They were helped by their central position between the old great powers of Egypt and Assyria and by the ample supplies of good ship-building timber from the Lebanon. They led the way in trade, in exploiting sea transport, and in the alphabet, which they invented and popularized wherever they went. But they remained, even in their most distant colonies, such as Carthage or Cadiz, too tied to the continuity of their culture with the old Babylonian civilization to do more than adapt it to the new conditions without generating much that was new, though we may suspect that what advances they did make were either destroyed or misappropriated by the Romans.

The Jews, closely related to them and sharing with them a mixed Egyptian and Babylonian culture, were reserved for a very different role in cultural history. Placed as they were at the very centre of warring peoples, Egyptians, Hittites, Philistines, and Assyrians, followed later by Persians and Greeks, and without the resources of overseas trade, their independence always remained precarious and was only saved in the end as a national entity by the evolution of a cultural tradition or law written in a book – the Bible. They were also, as a small people living in a relatively poor country, able to escape, though only by continuous efforts, from domination by native kings or oligarchs. For both these reasons independence, liberty, and democracy became indissolubly associated in their religion. In this the Jews were unique in the ancient world, and the influence of their religion and their sacred books was to prove of enormous importance to the subsequent development of civilization (pp. 225, 1035 f.).

THE BIBLE: LAW AND RIGHTEOUSNESS

The Hebrew Bible, or what we call the Old Testament, is far more than a collection of ancient history and legend, invaluable as that is for our

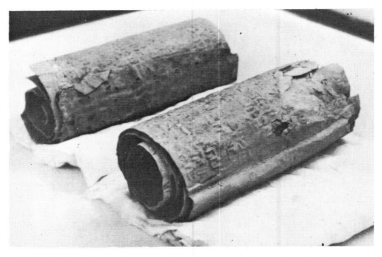

38. The Jewish religion, like some others and, in particular, the Christian, uses a sacred book. Recent research has made it appear that Jesus of Nazareth was a member of the Essenes, a sect closely linked with those who stored the early scriptural scrolls discovered in caves at Qumram, of which these copper scrolls of the Psalms are an example.

understanding of the past. It was first written down about the fifth century B.C. and has been preserved ever since as a religious and national rallying point. It is a book with a moral, full of propaganda expressed as poetry. Propaganda is as old as writing, but hitherto it had been the propaganda of the great and mighty, of the king and the priests. The propaganda of the Bible is different; it is essentially popular, stressing the ideas of *law* and of *righteousness*. Its unique character lay not in each of these ideas separately, because the Jews shared them with other cultures, but in their combination. Righteousness, as we find it in the Bible, is largely a protest against the abuses of the rich and powerful who, then as now, were addicted to falling into foreign ways of oppression. They could be restrained in the name of the law and the covenant with a timely backing of popular violence. The Jews were the first people we know of to fight for an idea and the wars of the Maccabees testified to their fanaticism and militancy. Jewish history is one of continual assertion of the people's right in the name of God. The Bible has, directly in Christianity and indirectly through the Koran in Islam, often served as the inspiration and justification of popular revolutionary movements (pp. 255, 1035).

GENESIS

It is, however, still another aspect of the Bible, one which is least characteristically Jewish, that has most affected science. The early books of the
Bible are versions of old Babylonian and even earlier Sumerian creation stories. They represent an attempt to account for the origin of the
world and man, which was an eminently creditable achievement in
3000 B.C., at the very dawn of civilization. These myths, once accepted
by the early Hebrew tribes, soon became the essential justification of the
covenant between God and His people and therefore beyond examination
and criticism. Later still, because they were part of the sacred books of
the Jews, these myths have come down to us as a literal divine revelation
to be accepted on faith.

Now the faith of the Jews, both in its original form and in that of a
Christianity largely derived from it, survived the break-up of classical
civilization because it was solidly based on popular feeling. It was thus
far better able to resist the stresses of bad times than the more logical
but scarcely more scientific constructions of the Greek philosophers,
which were felt to be by the common people, as indeed they were, elaborate justifications of upper-class rule.[2.74] In the new civilizations, which
grew out of the ruins of the old, religion was the central organizing principle, and accordingly the Bible and the Koran gained an absolute
authority in matters of science as well as in those of faith and morals.
The later chapters of this history will show with what difficulty and how
imperfectly human thought has managed to emancipate itself from these
fossilized relics of the myths of early man.

4.4 The Greeks

The most successful in the exploitation of the new conditions of the Iron
Age were the Greeks. They had the double advantage of being more
removed from the conservative influence of the older civilizations while
being able to make extensive use of their traditions. At the same time they
were protected in the early formative period of their culture by their
poverty, their remoteness, and their sea-power from the far less cultured
but more militaristic successors of the old empires – the land forces of the
Medes and the Persians.

The fact that the conscious and unbroken thread of history and science
comes to us almost entirely from the Greeks is an accident, but only

39. The Athenian Acropolis and Lycabettus Hill from the Hill of the Muses. As fortified in the sixth century B.C., it covered an area of about 300 metres by 145 metres, and was a sanctuary site of considerable antiquity, later associated with Athena. The Parthenon, here seen in ruins, was built *c.* 490 B.C. and is still one of the glories of Greek architecture.

partly an accident. The Greeks were the only people to take over, for the most part almost unconsciously and without acknowledgement, the bulk of the learning that was still available after several centuries of destructive warfare and comparative neglect in the ancient empires of Egypt and Babylonia. But they did far more than this. They took that knowledge and with their own acute interest and intelligence they transformed it into something at the same time simpler, more abstract, and more rational. From the time of the Greeks to the present day that thread of knowledge has never been broken. It may have been lost at times, but it has always been possible to find it again in time for it to be of use. The learning of the earlier civilizations has affected that of our own only through the Greeks. What we know now of the intellectual achievements of the ancient Egyptians and Babylonians from their own writings was learned too late to affect our civilization directly.

CLASSICAL CULTURE

In Greek lands there was built up between the twelfth and sixth centuries B.C. a unified culture which had made an ample digest of existing knowledge and added to it far more of its own. The resulting *classical* culture, as we now call it, enlarged but not seriously modified by that of Alexandria and Rome, has remained the essential corner-stone of our modern world culture. Classical culture was synthetic; it made use of every element of culture which it could find in the countries it occupied and with which it came into contact. It was not, however, a mere continuation of these cultures. It was something definitely new. The characteristics of classical culture that distinguish it are not, however, any of those which are sometimes called cultural. There have been other civilizations, both before and after, that have had as distinctive an art and literature. The great contributions of classical culture were in political institutions, particularly democracy, and in natural science, especially mathematics and astronomy.

THE BIRTH OF ABSTRACT SCIENCE

The unique character of Greek thought and action resides in just that aspect of their life which we have called the scientific mode. By this I do not mean simply the knowledge or practice of science but the capacity to separate factual and verifiable from emotional and traditional statements. In this characteristic mode we can distinguish two aspects: that of rationality and that of realism; that is, the ability to sustain by argument and the appeal to common experience.

That the Greeks could achieve this, even partially, is due to the historical circumstances in which their culture took shape. The Greeks did not make civilization or even inherit it – they discovered it. The enormous advantage they gained by this was that for them civilization was something new and exciting and could not be taken for granted. The original culture of the mainland Greeks was of the simple European peasant type. It was unable to stand up against the much more elaborate cultures of the countries the Greeks moved into – that extremely rich and mysterious secondary culture of Crete and Anatolia from which so much of classical culture is derived. Words ending in 'issos' and 'inthos' seem to be of Cretan origin; some have passed on to us, as in the names of Narcissus and Hyacinth. The influence on the Greeks of the original centres of civilization, Mesopotamia and Egypt, was not to arise till much later.*2.16

In losing their original culture, however, the Greeks did not and could not take over the cultures of the other countries in their entirety. What they did was to select from foreign cultures what seemed to them to matter. This included in practice every useful technique and in the field of ideas mainly the explanations of the workings of the universe, rejecting the enormously complicated elaboration of theology and superstition that had been built on them in the decadent period before and during the iron age invasions. Homer, the first and greatest of the Greek poets, fixed for all time the picture of the world the Greeks came into. In the *Iliad* and *Odyssey* we find an enormous contrast between the simple peasant life of the newly arrived Hellenic clans and the complex rich and ancient civilizations that they discovered only to destroy. Homer's poems remained as the bible of the Greeks, providing the common basis of belief about gods and men and the arts of peace and war. They contained as much science as the average man ever needed to know.

THE ECONOMIC BASIS OF THE GREEK CITY

Greek culture, in common with most of the Western iron age cultures, had such a different economic base from that of the older river-irrigation cultures that much of their way of life was intrinsically unassimilable. It depended on a rather poor kind of dry farming with small peasant holdings helped out by vineyards, olive groves, and fishing. Hesiod, a poet of the early Greek period, describes this life in rather grim terms. His father's land at Ascra in Boeotia he describes as 'cold in winter, hot in summer, good at no time.' Nevertheless, though liable to periodic debt crises, iron age economy was, until slavery was introduced on a large scale, basically stable. It was supplemented and balanced by extensive foreign trade, no longer as in the older civilizations mainly in luxuries for temples and palaces, but in bulk commodities for the common citizen.

The most characteristic Greek city state of Attica was so short of good corn-growing land that it depended on its exports of pottery, olive oil, and silver to buy the food for the relatively enormous population of over 300,000 of the city of Athens. The early Greeks were able to exploit their local resources to the full with all the intensity and simplicity that are possible only in a compact city. In these circumstances there were rapid and even violent economic and political changes, while tradition, though never lost, was at a discount. The more enterprising citizens had both the incentive and the ability to think out what they wanted to do and to do it. In the measure that they succeeded they could improve their status in society, held back neither by clan nor by state barriers. Institutions and divinities became less important and more attention was concentrated on men.

ART AND DIALECTIC

The realistic representation of man in painting and sculpture, in drama and in science, was the characteristic new feature of Greek civilization. Greek art as represented on statues and vase-paintings – the large frescoes have all been destroyed – shows a concentration on the naked human body that would seem odd if we were not so accustomed to it. It derives originally from ritual games and the cult of athletics that sprang from them. Egyptian statues had a directly magical purpose: they had to re-embody the spirit (Ka) of the dead man, they had to be lifelike to work. The Greek sculptor was more sophisticated. He was trying to suggest an ideal to be aimed at in human bodily perfection. In Greek culture the athlete, the artist, and the doctor worked closely together, and this resulted, among other things, in a concern of the medical profession with health rather than disease.

Realism in art went with rationality in words. Because of the fading of ancient sanctions each case had to be argued out on its merits. The history of Greek philosophy and Greek science – the two were never distinguished in those days – is the history of such a sequence of arguments; back-and-forth arguments of the kind that they called a *dialectic*. The capacity to argue was also made possible by the political features of Greek life. The small city state gave much more scope for the average *individual* citizen than did the capital of a large empire. At the same time the intense political life of the city, with its emphasis on trade deals and lawsuits, in which at first every man was his own lawyer and judges were chosen by lot, made it possible and indeed necessary to develop argumentation to the highest

40 a, b. The Greek and Egyptian approaches to the human form were very different. In spite of the formal headdress and the ritualistic attitude of the mummified body, the Egyptian sculptor showed life as it was seen to be. The wooden statue (a) was found in a funerary chapel of a fifth dynasty tomb at Saqquara (*c.* 2450 B.C.), and it has an astoundingly lifelike quality. It was nicknamed the 'village sheikh' by Auguste Mariette and his workmen. It is now in the Cairo Museum. As a contrast, the Greek statue (b) of Hermes by Praxiteles (*c.* 364 B.C.) has a formality of idealization of the male figure that stresses the aesthetic ideal rather than a realistic attitude.

degree. This emphasis on the mastery of words led to a great literature and oratory, but it had the disadvantage of drawing thought away from the study and handling of things.

THE SEPARATION OF SCIENCE FROM TECHNIQUE

Greek science has an altogether different character from that of the early civilizations; it is far more rational and *abstract*, but it remained as far or farther removed from technical considerations. Its traditional presentation is in the form of an argument based on general principles, rather than as examples drawn from the particular problems of technique or administration such as we find in Egyptian or Mesopotamian texts (pp. 120 f.). Mathematics, especially geometry, was the field which the Greeks esteemed most highly and where their methods of deduction and proof are those we still use. Because of the immense prestige of these methods we are apt to overlook the fact that they are applicable to only a very limited part of Nature, and even there only where the spade-work of observation and experiment has been done. A belief that the universe is rational, and that its details can be deduced from first principles by pure logic, certainly served in the early days of Greek science to liberate men from superstitions. Later, particularly after Aristotle had become an authority instead of, as he wanted to be, an instigator of research, this abstract and *a priori* approach was to prove disastrous to science. It led generations of intelligent people into the belief that they had solved problems which they had not even begun to examine (pp. 297, 1036).

The technical developments made in the early Iron Age, and especially by the Greeks before the Alexandrian period, though important in their effects, were not innovations as fundamental as those of the Bronze Age. The use of iron led directly to the improvement of all hafted tools such as axes and hammers, and also made possible implements like the *spade* which would have been too expensive in bronze to be of any use. It probably also made possible the use of the *hinge*, leading to two new tools of some importance – the tongs and the drawing compass. These all arise from the ease with which iron bars may be bent over in a loop and then welded to form a hole for a haft or peg. It was not so much the improvement of the tools as their ready availability that constituted the revolutionary technical advance of the Iron Age. Later, through the marriage of Greek mathematics and Egyptian or Syrian techniques, came the most important developments, including, as we shall see (p. 218), a whole host of applications of rotary motion, mills and presses, pulleys and windlasses, as well as hydraulic and pneumatic devices, water-lifts and pumps.

Of chemical inventions the most important is that of blown glass first made in Egypt, though this remained for long a luxury product. As a result of a few innovations and many improvements, the efficiency of classical techniques, particularly metal-using techniques, was by the sixth century B.C. well above that of the bronze age cultures in their heyday. This was one of the reasons why the Greek armoured soldiers were for a few centuries able to overcome far larger numbers of Asiatic troops.

The technical advances of the Iron Age did not, however, affect the learned in the same way as had those of the early Bronze Age. It was partly because they were essentially improvements and not radical innovations that they did not strike the imagination. Further, they created little demand for new auxiliary scientific techniques. There was enough arithmetic and geometry to cope with them already. The most

41. Low-relief carving of Asclepius, the Greek god of healing, *c.* fourth century B.C. The cult of Asclepius appealed strongly to the rising individualism of this period. The Asclepiañ cures contained the ritual of incubation, auto-suggestion, dietetics, baths – those at Pergamum are known to have had radioactive springs – and exercise; the sanctuaries were, in fact, sanatoria. Now in the Altertumsmuseum, Mainz.

powerful reason, however, was that the craftsman was still despised. The hand worker, *cheir ourgos* in Greek (our surgeons are still called Mr, not Dr), was considered a definitely inferior being to the brain worker or contemplative thinker. This was no new idea; it was inherited from the old civilization (pp. 117, 130), but it was strongly reinforced, especially in later Greek society, by its association with slavery. Although much craft work was done by free men they were degraded by competition with slaves, so that their work was called base or servile!

In the same way a slave society debased the economic and social position of women. Indeed, the position of the wives and daughters of Greek citizens was far worse than it was in the older civilizations. They were precluded from taking part in public life and were little better than domestic slaves. As a result all domestic work, which included far more arts than it does now, such as weaving and the preparation of simple remedies, was beneath the concern of the philosopher. For although the philosophers drew on the work of the craftsmen for the derivation of their ideas as to how Nature worked, they had little first-hand acquaintance with it, were not called on to improve it, and consequently they were unable to draw from it that wealth of problems and suggestions that was to create modern science in Renaissance times.

ARCHITECTURE

There is one important exception to be made to the general contempt for mechanical operation. Architecture in Greek times advanced to the level of a citizen's profession, not of a mere manual art. We all know the triumphs of beauty, proportion, and symmetry of Greek architecture and the impressiveness of the Roman architecture that followed it. Now architecture is pre-eminently an art depending on geometry and involves accurate drawing. It could therefore hardly fail to affect the queen of Greek science, mathematics. Two instruments helped in the same direction, the draughtsman's compass and the lathe. The compass was such a convenient, accurate tool that it is not surprising that Greek geometry tied itself almost exclusively to ruler-and-compass constructions. The pole lathe, with its backward and forward motion, derived from the bow drill, was a bronze age invention; the modern lathe, belt-driven, came only in the fourteenth century A.D.,[2.54] though pole lathes are still in use in many parts of the world, and were in England down to fifty years ago. On the lathe it was possible to turn cylinders, cones, and spheres, and they provided admirable playthings for the mathematician. The degree to which the techniques influenced science in Greece was not negligible, but it was relatively far less than in the older civilizations. Greek science accordingly developed in a more general and independent way but, lacking the check of experience, it was apt to get lost in guesses and abstractions.

CONTENT AND METHOD IN GREEK SCIENCE

Nevertheless, modern science is directly derived from Greek science, which provided it with an outline, a method, and a language. All the general problems from which modern science grew – the nature of the heavens, or man's body, or the workings of the universe – were formulated by the Greeks. Unfortunately, they also thought that they had solved them in their own particularly logical, beautiful, and final way. The first task of modern science after the Renaissance was to show that for the most part these solutions were meaningless or wrong. As this process took the best part of 1,400 years it might be argued that Greek

42. The pole lathe was a bronze age invention, driven by a bow and treadle. Although no illustrations of a bronze age lathe of this kind appear to exist, even after the development of the modern belt-driven lathe some time in the fourteenth century A.D., the pole lathe was not ousted, as this copperplate engraving from the second edition of Jacques Besson's *Theatre des Instrumens Mathematiques et Mechaniques*, Lyon, 1593, shows (first edition 1579).

science was a hindrance rather than a help. However, we cannot tell whether, in the absence of Greek science, the problems would have been set at all.*

STAGES IN THE DEVELOPMENT OF GREEK SCIENCE

The history of Greek science, though it formed one continuous movement, may conveniently be split into four major phases, which may be called: the Ionian, the Athenian, the Alexandrian or Hellenistic, and the Roman phases. The Ionian phase (4.5) covers the sixth century B.C. and is that of the birth of Greek science in the region where the influence of the older civilization was most felt. It is associated with the legendary figures of Thales and Pythagoras and other Nature philosophers who speculated, in a most materialistic way, on what the world was made of and how it had come to be. This philosophy, as becomes an age of social development, was essentially positive and hopeful.

The second phase (4.6) covers the years from 480 to 330 B.C., between the successful end of the Persian wars and the effective suppression of the independence of Greek cities by Alexander the Great. It was during this period that Greek culture reached its peak of achievement in the Athenian democracy of the age of Pericles, only to destroy itself in civil strife and war. In this period the interests of philosophy shifted from the explanation of the material world to that of the nature of man and his social duties. This was the great period of Socrates, Plato, and Aristotle, usually considered as the high point of Greek wisdom.

The third phase (4.7) of Greek culture, that called Hellenistic, began with the decadence of the independent city states and their supersession by land empires of a new type. The empire of Alexander brought Greek science once again into direct contact with the older sources of culture in the East as far as India. Alexandria became a new home for science where, for the first time in history, it was subsidized through the founding of the Museum. The result was the great development of mathematics, mechanics, and astronomy that we associate with Euclid, Archimedes, and Hipparchus. In the history of science, as distinct from philosophy, this third phase was to be the most important of all, for it was then that the body of exact science was first formed as a coherent whole, and enough of it survived despite the losses of the dark ages that followed to set science going again nearly 2,000 years later. From the second century onwards, with the coming of the Romans, this effort slackened and came to a stop long before the actual fall of the Empire. This last phase (4.8) cannot be distinguished by any originality, but as it was to be the bridge between classical and all later science it deserves separate consideration.

43. Greek armour had few variations. The armour of the hoplite has been mentioned above (36), and the Greeks never developed cavalry armour to any great degree since the cavalry was composed of a few nobles. Indeed, in the attempt of Cyrus the Younger in 401 B.C. to take the Persian Empire from his brother Artaxerxes, it was the smallness of his cavalry force that was the decisive factor in his defeat. The photograph here is of an early fifth-century Greek cavalryman and, except for the helmet, the rider appears to have little, if any, protection. Only in its Hellenistic phase did the development of cavalry with leather or metal armour occur. Now in the British Museum.

4.5 Early Greek Science

IONIAN NATURALISM

Greek science is usually recognized as originating in the Ionian cities of Asia Minor, particularly Miletus, where contact with the ancient civilizations was closest, and in the new colonies of Greeks that had been formed in Italy and Sicily. It appeared in the sixth century B.C., just at the time when the rule of the old landed aristocracy was breaking up and power was being seized by a set of local bosses, the *tyrants*, with the support of the trading classes. The Greek world of the sixth century was one of violent expansion. Its commercial centre was first the eastern Aegean, settled mostly by Ionians, one of the original tribal groups of the mainland Greeks. These set up colonies over the Mediterranean as far as Marseille, Naples, and Sicily, and east over the coasts of the Black Sea. When Persian pressure drove them from their original homes the colonies in turn became centres of trade and culture of essentially the same character. That is why it is reasonable to include Thales from the mother city Miletus, Heraclitus of near-by Ephesus, Pythagoras a refugee from Samos settling in south Italy, and Empedocles of Sicily all in the group of Ionian philosophers.

In this time and environment tradition was at a discount and new answers to old questions had a chance to be heard. The great value of the early period of Greek thought was that it tried to answer all questions in a simple and concrete way. It was an attempt to formulate a theory of the world – what it is made out of and how it works – in terms of ordinary life and labour.

PHILOSOPHERS AND SAGES

The people who asked and answered these questions were only later, by Socrates, called *philosophers*, i.e. lovers of wisdom. In their own time they were called sophists, i.e. wise men. We now know very little about them or what they believed; most was handed down by oral tradition and finally a few fragments have been saved by references in the works of Plato and Aristotle, who used them chiefly to refute or make fun of their predecessors. The very fact that they were known and remembered and that legends about their lives have persisted shows how important they must have been in their time. When a new civilization was crystallizing after the warfare of the early Iron Age these philosophers presented a new social type. Nevertheless they were in essence wise men or sages who had picked up and were retailing the old knowledge of the East, adapted and improved on to suit the new times. They were also prophets

44. The Erechtheum, often spoken of as the most notable Greek Ionic building in existence. It was constructed 421–407 B.C. of Pentelic marble with friezes of black stone to take white marble relief sculpture. At the west end is a south-projecting feature usually known as the 'porch of maidens' with caryatids acting as supports. The beauty of design and detail in the building is what makes it so notable. Its name is derived from Erechtheus, the fabulous king of Athens who was reared by Athena.

and leaders of religious mysteries, often founding semi-monastic communities that were also schools. Those that succeeded – and they are the only ones that we hear about – usually managed to secure the position of political or scientific adviser to some tyrant or democratic boss, and were consulted or probably gave gratuitous advice on every kind of subject. If they quarrelled with their patron they were usually snapped up by a rival. It added prestige and stability to a government to have a famous philosopher behind it. Pericles for instance had the benefit of the presence of Anaxagoras, but this time the philosopher went too far in flouting popular beliefs and he had to be dropped. Whether they favoured the democratic or aristocratic side, they were nearly all well-to-do gentlemen. We hear of a few who had to work for a living; Protagoras and other sophists of the fifth century accepted fees for teaching. Plato, who was rich enough not to need to, sneers at them for doing so. They were, he felt, losing their amateur status as philosophers.

It was not only in Greece that such philosophers were to be found. In many parts of the world the disturbances of the Iron Age gave scope to men with similar ideas and messages. In Palestine there were the prophets and the later authors of the Wisdom literature such as Ecclesiastes or the book of Job. Jeremiah may well have met Thales at Naucratis in Egypt. In India there were the rishis and buddhas, of whom Gautama, the Buddha, is the most famous. In China Lao-tze and Confucius lived at about the same time. All had in common the formulation of general views of the world of Nature and man. Most advised princes and tried to reform States without any lasting success. Most were unorthodox in their time, even when they claimed, as Confucius did, that they were trying to recapture the wisdom of the Ancients. It was only later that they were to become the founders of new orthodoxies.

Their success was due to the fact that they filled the gap in ideas left by the economic transformation of a bronze to an iron age civilization. They provided what Marx called the ideological superstructure for a new system of relations of production. In that new system the direction of society in the hands of merchants, tyrants, and military princes was apparently more divorced from the material side of production than it was in the Bronze Age. Nor did the philosophers, unlike the great directors of works of the time of the canals, pyramids and temples, have anything to do with the actual material running of the economy. As a result the superstructure they put up was, in general, idealist and inimical to the growth of experimental science.

The early Ionian philosophers do not, however, fit entirely into this picture. In their time the slave State and the rule of the rich were by no means fully established. Accordingly they differed from most of the wise men of the East in that they were at the same time materialistic, rational, and atheistical. They were concerned less with morals and politics and more with Nature than their successors.

THE WORLD AND ITS ELEMENTS:
THALES, HERACLITUS, AND EMPEDOCLES

The first of the traditional Greek philosophers was Thales. He is credited with holding the theory that everything was originally water from which earth, air, and living things separated out. This is recognizably the same theory as that of the book of Genesis, a common Sumerian creation myth, reasonable enough for a delta country where the dry land had to be won from the marshes. These myths, because they were faithfully preserved in the original form which dates from before the first class societies, are fundamentally materialist.[2.85] What is new about Thales' version of these is that he left the creator out. Like Laplace centuries later in his answer to Napoleon, 'he had no need for that hypothesis'. The materialism of Thales is contained in his interest in Nature and rejection of metaphysical speculation, which was later interposed to justify a class society. It is not a mechanical materialism but rather one in which all matter is thought of as alive. He was a *hylozoist* (matter-life). This basic materialism and atheism was maintained by the later philosophers of the school, Anaximander and Anaximenes, who modified the hypothesis to make it account for more phenomena. They also invoked earth, mist, and fire, as the *elements* (l, m, n) (in Greek – *stoicheia* or letters) out of which the world was made, as words are spelled out of letters. Heraclitus, the philosopher of change, took as his motto *panta rhei*, everything flows. He thought fire was the prime element because it was so active and could transform everything. His expression for this is revealing: 'all things are an exchange for fire, and fire for all things, even as wares for gold, and gold for wares.'[1.29.16] This reveals once again how technical processes and economic practice engendered the new philosophy. He also introduced the idea of opposites, some things, like flame, tending to move up, and others, like stone, to move down. The two opposites were necessary to each other and generated a tension like the bow and its string. This is the first enunciation of a dialectical philosophy.

Empedocles, the successor of this school of materialist philosophers, demonstrated by experiment that invisible air was also a material substance, and fixed the order of the ancient elements as earth, water, air, and fire, one above the other, each striving if disturbed to get back to its place. He thought that the opposite tendencies, love and hate, which

45. The geocentric universe was universally accepted in ancient times. It did not remain unquestioned, but such objections as did arise were overruled for aesthetic and philosophical reasons. The theory, with slight modifications and, most particularly, accompanied by a firm belief in the existence of a transparent crystal sphere for each planet, was current in western Europe, and the illustration here is taken from Peter Apian's *Cosmographia*, Antwerp, 1539, which was edited and published posthumously by his pupil Gemma Frisius. It shows the Earth with the four Aristotelian elements, earth, air, fire and water, in the centre, and above them the celestial spheres. The first is that of the Moon, followed by those of Mercury, Venus, the Sun, Mars, Jupiter and Saturn, the outermost planet known until well after the invention of the telescope. Then comes the sphere of the fixed stars; beyond, that of the ninth sphere (a necessary fiction to account for the observations of the apparent motion of the stars due, as we now know, to the Earth's motion in space) and then Aristotle's 'Primum Mobile' – the tenth sphere driven by the divinity and from which all other spheres derive their motion. Beyond this lay heaven, 'the habitation of God and the Elect'.

he also conceived as material principles acting mechanically, were continually mixing the elements up and separating them out again. This is similar to but probably quite independent of the Yin and Yang dualism of ancient China. Here also we have two principles, male and female, fire and water, interacting to form the remaining elements, metal, wood, and lastly earth, and from them by further mixtures the 'ten thousand things' of the material world.

The whole trend of Ionian thought was towards a dynamic world of *continuous mutual transformation* of *material elements*. Most philosophers of later times tended to concentrate more on the static *natural order* of the elements and to think of them as a fixed and unalterable part of the structure of the universe. This static order of elements consecrated by Aristotle was used to limit any kind of progressive change, particularly social change. This could be done by equating elements to social classes and inferring that the ideal and final state of the social universe was one in which the lower classes were subordinated to the upper. The identification of the social and natural worlds hindered the understanding of either. It turned an originally materialist theory into a formal one and hindered the development of astronomy, medicine, and chemistry by saddling them with far-fetched analogies which claimed the sanction of a universal order.

Another deep-seated confusion underlay the world view of the Ancients – their elements had to fulfil two incompatible functions. In one aspect they stood as the actual materials and motions of the world as they knew it; they served to explain without any recourse to the gods the whole panorama of land and sea, sunshine and storm. In that sense we still speak of the fury of the elements. In quite another way the elements also stood for qualities – hotness and coldness, wetness and dryness, lightness and heaviness – attributable to anything. Each element was not pinned down to a particular material substance, as were the chemical elements of the nineteenth century. Anaxagoras (*c.* 500–428 B.C.), last of the Ionians, went so far as to say that the seeds of every element were in everything, like our present *states of matter*, gaseous, liquid, and solid.

The triumph of the original Ionian school was that it had set up a picture of how the universe had come into being and how it worked without the intervention of gods or design. Its basic weakness was its vagueness and purely descriptive and qualitative character. By itself it could lead nowhere; nothing concrete could be done with it. What was needed was the introduction of *number* and *quantity* into philosophy.

QUANTITY AND NUMBER : PYTHAGORAS

The tendency to associate arbitrary simple number ratios with celestial objects, which may well have had its origin in Babylonian astronomy, already appeared in the work of Anaximander (611-547 B.C.), who put the distance of the stars, moon, and sun at nine, eighteen, and twenty-seven times respectively the thickness of the earth disc. The attribution of *numbers* to all aspects of Nature is associated with the doctrines of Pythagoras (582-500 B.C.) He came from Samos, an island near Miletus, but emigrated to south Italy, where he founded a kind of philosophic, religious school. Whether Pythagoras was an entirely legendary figure or not, the school that bore his name was real enough and was to have an enormous influence in later times, particularly through its most important exponent, Plato (427-347 B.C.).

Two trends of ideas are blended in Pythagorean teaching, the *mathematical* and the *mystical*. It is doubtful how much of the Pythagorean mathematics was his own. Certainly his famous theorem on the right-angled triangle had been well known as a practical rule to Egyptians, and the Babylonians made long tables of 'Pythagorean' triangles. It may even be that the whole of Pythagorean number theories, in their mystical as well as their mathematical aspect, are drawn from some source in Eastern thought, as their character strongly suggests. But whether Pythagoras was an originator or a transmitter, the connexion set up by his school between mathematics, science and philosophy was never again lost.

Pythagoras saw in *numbers* the key to understanding the universe. He related them on the one side to geometry, showing how squares and triangles could be made of appropriately arranged points, and on the other to physics with the discovery that strings which were in simple *ratios* of length emitted notes with regular musical intervals – octaves, thirds, etc. – between them. This linked the previously sensuously appreciated *harmony* with ratios of numbers and hence with geometrical forms. The Pythagoreans set the whole tone of Greek geometry by their insistence on the cosmic importance of the five regular solids whose sides could be made from triangles, squares, and pentagons. The pentagon was particularly magical because its construction with ruler and compass was a mathematical triumph. Two of the Platonic solids; the dodecahedron and the icosahedron, have pentagonal symmetry: Euclid's whole geometrical synthesis indeed leads up to the method of construction of these two solids; and the proof that there can be no more was a culminating point of Greek geometry, foreshadowing the modern theory of groups.*

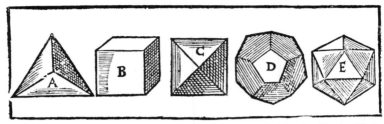

46. The five regular geometrical solids were much studied by the Greeks. Their symmetry and the progression from one to another had an aesthetic as well as a mathematical appeal and they were ideal subjects for the geometer. They are: A the tetrahedron, each face formed by an equilateral triangle, B the cube, C the octahedron, D the icosahedron and E the dodecahedron; in all the faces are each equal in area and in shape. Pythagoras or the Pythagoreans are traditionally credited with the discovery. From Levinus Hulsius, *Instrumentorum Mechanicorum*, 1604.

RATIO AND IRRATIONALS

One fundamental mathematic discovery came from the Pythagorean school, though probably some time after the death of the master. If every measure of length can be expressed by a number, the *proportion* between two different measures should be expressible as the *ratio* of two numbers. But a very simple case shows that this cannot be done. Whatever number you use to express the length of the sides of a square, that of its diagonal cannot be expressed as another number, whole or fractional. This is equivalent to saying that no fraction multiplied by itself can give exactly 2, or that $\sqrt{2}$ is *irrational*. The discovery that there were irrational numbers was a serious shock to the whole Pythagorean school and contributed to its break-up. One way out was to say that measures were unreal; the other, that finally adopted, was to extend the concept of numbers to include irrationals.[2.72]

It is to the Pythagoreans that we owe the importance of the circle and the sphere in astronomy. They thought that earth was a sphere and further that it moved together with the planets – the sun, moon, and a mysterious counter-earth – round a permanently invisible central fire. This idea, when rationalized by Heraclides (375 B.C.) and Aristarchus (c. 310–230 B.C.), was to lead to the modern picture of the solar system.

The work of the Pythagorean school is the very foundation of mathematics as well as of the physical sciences. Even in mathematics the mystical element is very much in evidence. The Pythagoreans linked the eternal soul with the eternal forms of number, attributing it particularly to the number $10 = 1 + 2 + 3 + 4$. The whole world, according to them, was made of pure numbers. This form of extreme idealism is linked with cabbalistic number magic, still invoked in the blessed trinity, the four evangelists, the seven deadly sins, and the number of the beast. It is also apparent in modern mathematical physics whenever its adepts try to make God the supreme mathematician.

MYSTICISM ENTERS SCIENCE

In physics also the Pythagoreans ran too much beyond the facts, and substituted number mysticism for experimental knowledge. The mystical side of Pythagoreanism links it with the Orphic mysteries, a relic of old community magic that had already become a means of escape from the harsh realities of the Iron Age.[2.82.154] Orphism as a slave religion has indeed some points of resemblance to Christianity, especially in its symbolism of the wheel and the cave.[2.84] The main thesis of the Pythagoreans was the doctrine of the transmigration of souls, essentially the same as that of the Hindus, though possibly quite independent of Indian influence. The object of the cult is to escape from the cycle of reincarnation by common mystical experiences, '*orgies*', and ecstatic mystical contemplation, '*theories*' = visions.[2.24.38] This is similar to the idea of the attainment of Nirvana through Yoga which Gautama tried vainly to resist. The idea of rebirth was not unreasonable in the Old Stone Age, where it first appeared. In the Iron Age it was essentially reactionary because it removes all meaning from social injustice and war, and wins for them at least a tacit approval (p. 1035).[2.20] In the *Bhagavad Gita*, when Arjuna asks in horror how fractricidal strife came about, Krishna answers:

> If the red slayer thinks he slays
> And that the slain thinks he is slain,
> They little know the hidden ways,
> I turn, I pass, I come again.

The mystic aim was the achieving of detachment through purification. This purification was originally a purely magical initiation ceremony or rebirth. It later acquired a link with alchemy through the purification of metals by fire (p. 127). The Pythagoreans introduced the idea of purification through knowledge, the *pure knowledge* of passive *contemplation*. This view was that the people, like spectators at the games, can be divided into three classes: those who go to buy and sell, the competitors, and the spectators.[2.15] The last, who merely contemplate, the Pythagoreans deem far superior. This ideal of pure science as contemplation, drawn from a primitive ritual debased by class society, has lasted down to our time.

47. The triangle, the five-pointed star or pentagram, and more complex star-shaped figures were all part of the mysticism that permeated Pythagorean teaching (see page 179). This mysticism continued to be attached to geometrical shapes, especially by the alchemist. The tree of universal matter with the Sun (gold), the moon (silver), and the star shapes containing (left to right): Mars (the symbol for iron), Venus (copper), Mercury (quicksilver), Saturn (lead) and Jupiter (tin). In the lower triangle are the signs for the sulphur–salt–mercury theory of primary matter. From *Occultal Philosophia* . . . , Frankfort, 1613. The names Senior and Adolphus refer to the two semi-allegorical human figures.

Now, as then, it provides a convenient excuse for enjoying knowledge without responsibility.

Though these consequences of Pythagorean views are clearly reactionary they come from an age later than that of Pythagoras himself. The original Pythagorean community, according to Thomson,[2.84] was as much political as religious, and as such was persecuted and finally dispersed. He regards Pythagoreanism as the first expression of *democratic* thought, that is of the rationalism of the merchant *middle* classes as against the traditionalism of the landed aristocracy, and compares its influence with that of Calvinism. In particular he links the Pythagorean insistence on the value of the *mean* and of harmony with the solution of the political struggle through the rise of the merchants, an idea which we now associate with Aristotle.

THE INFLUENCE OF PYTHAGORAS

The school of Pythagoras marked a branch point in the development of Greek science both in theory and practice. From it stem two very different systems of thought. The most abstract and logical aspects were taken up by Parmenides and, mixed with much mysticism, became the basis of Plato's idealism. In the opposite direction Pythagoras' number theory was given a materialist content in the atomic theory of Leucippus of Miletus, 475 B.C., and Democritus of Abdera, 420 B.C.

In practical science the Pythagoreans established the possibility of dealing with physical quantities by reducing them to measure and number, a general method which, though often stretched beyond its proper limits, was to provide one continuous means of expanding the control of man over Nature. For mathematics the importance of Pythagoras was even greater in that his school established the method of *proof* by *deductive* reasoning from *postulates*. This is the most powerful way of *generalizing* experience, as it transforms a number of instances into a *theorem.** Valuable as it is in mathematics, deductive proof has been used ever since in the service of idealism to prove palpable nonsense from self-evident principles.

PARMENIDES

Among the first philosophers to do this were Parmenides (470 B.C.) of Elea in south Italy and his pupil Zeno (450 B.C.), both connected with the aristocratic and conservative party in the city. Parmenides was the philosopher of pure reason. He violently attacked the whole of observational and experimental science, claiming that such studies could only give uncertain opinions owing to the fallibility of the senses, while the truths

48. An arrow shooting machine, from an early seventeenth-century copperplate engraving. Machines of this kind were in use in Greek times. The flight of an arrow was one of the matters dealt with by Zeno who discussed whether space is continuous or discontinuous, and proved that in either case an arrow cannot reach its goal – nor can a runner for that matter. From Justus Lipsius, *Poliorceticon*, Antwerp, 1605.

of number appreciated by pure reason were absolute. The demand for *absolute truth* and certainty, not to be found in the fallible senses, in 'the blind eye, the echoing ear', expresses the deep need for fixity that always recurs, usually on the losing side, in times of trouble.

It is not surprising that this anti-scientific idealistic trend was later taken by Plato and has persisted in philosophy to this day. Parmenides went further; he refuted, by appealing to logic, Heraclitus' view that everything changes. If *what is, is* and *what is not, is not*, nothing can ever happen, and *change is impossible*. Not only change but also variety is impossible in such a universe. The *real* universe is one and changeless. As our senses show us variety and change they must be only seeming, and the apparent material world must be an *illusion*. This is the first clear statement of the extreme idealistic view, and the beginning of *formal logic*. Hegel took up Parmenides' logic, and refuted his proofs by claiming that the idea of being contradicted by the idea of not being gives rise to the idea of becoming, and thence by the same *dialectic idealism* to the whole complex ideal world. This was the philosophy which Marx turned on to its feet in founding *dialectic materialism* (pp. 1100 ff.). Parmenides' idealism is an extremely convenient one for a minority ruling by 'divine' right.

Parmenides' pupil, Zeno, attacked the basis of Pythagoras' mathematical and physical theory by producing four ingenious paradoxes which appear to prove logically that time or distance can neither be continuous nor discontinuous. If space is continuous the runner can never reach the goal. If he is halfway it will take him time to get half the rest of the way and so on *ad infinitum*. If space is discontinuous the arrow can never move because it is either at one point or the next and there is nothing between. Zeno's paradoxes were not entirely useless – they are the beginning of the search for *rigour* in mathematics. These subtleties were taken to prove that the visible world cannot really exist; but they may serve as well to show that pure reason can be sillier and emptier than anything the senses can contrive.

ATOMS AND THE VOID : DEMOCRITUS

The most effective answer to these idealist tendencies was given by Democritus, whose *atomic theory* was to have such an enormous influence on later science. Instead of thinking of a universe of ideal numbers he imagined one made out of small innumerable uncuttable (a-tomos) particles, *atoms* moving in the *void* of empty space. The atoms were unalterable, to this degree agreeing with the changelessness of Parmenides; they were of various geometrical forms, to explain their capacity for combining to form all the different things in the world; and their *movement* accounted for all visible change. Thus Democritus was able to include the mathematical content of Pythagoras, especially the insistence on the importance of geometrical form, while rejecting its idealism and mysticism.

The introduction of the void - nothingness – into philosophy was also a daring step. The universe of the older philosophers was that of common sense; it was a full universe, a plenum. The idea of a *vacuum* was abhorred by all reputable philosophers, and the abhorrence was fathered on to Nature. Many of the great achievements of Renaissance physics, like Galileo's dynamics, and later scientific and technical developments, such as the laws of gases and the steam-engine, arose in the process of overthrowing this idea (pp. 469 f.).

The atomic theory had from the start a radical political flavour because it was frankly materialistic and avoided appeal to preordained harmonies. The authority of Plato and Aristotle, who supported doctrines of ideals or substantial forms (pp. 204 f.), was sufficient to prevent its general acceptance. Nevertheless it remained throughout the classical period as a persistent heresy, and through Epicurus and Lucretius had an effect on philosophy and ethics in its later stages. It stood for a world which maintained itself through the natural working of its parts and needed no divine guidance. The atomism of Democritus was completely deterministic, but later Epicurus introduced a certain amount of original variation or bias to his atoms in order to allow for variety and for free will in man.[2.60]

It would be a mistake to think of Greek atomism as essentially a scientific theory of physics. No conclusions were drawn from it which could be practically verified. Nevertheless, it was the lineal and acknowledged ancestor of all modern atomic theories.[2.93] Gassendi (p. 463), the first of the modern atomists, drew his ideas straight from Democritus and Epicurus. Newton (p. 466) in his turn was a fervent atomist, and it was the inspiration of his work that finally led John Dalton (p. 626) to found the atomic theory of chemistry. The atoms of chemistry have not proved as uncuttable as their name implies, but the deeper explanations of nuclear physics still lie in the same atomic tradition.

THE AGE OF PERICLES

The city of Athens emerged at the end of the Persian wars in 479 B.C. as the economic and cultural leader of the Greek world. It had earned that place by its courage and persistence in opposing the invader. Its success was, in fact, largely due to the use to which it put the money drawn from the Laurion silver mines. On Themistocles' advice it went to build up a navy which, manned by the poorer citizens, ensured not only victory for the city but also the power of the common people in its government. The commercial leadership of Athens still further increased its wealth and drew to the city not only artists and sculptors but also historians and philosophers. For the next century, even after the disastrous war with Sparta, Athens was the intellectual centre of the Greek world; and the heritage of Ionian science, particularly the mathematical, astronomical Pythagorean tradition, there received a new impetus.

The period is one of enormous importance to the development of world science because it furnishes the link between the poetic speculations of the Ionians and the precise calculations of the Alexandrian period. Indeed, the last of the Ionian philosophers, Anaxagoras of

Clazomenae, settled in Athens, was the friend of Pericles, and was, as has been told, expelled for rationalism in 432 B.C.

It was in this period that the major problems of science, social as well as natural, were set, though many diverse solutions to them were to be proposed in succeeding centuries. From then on Greek science was to be autonomous and to develop its own particular character within its own, largely unrealized, limitations. In the natural sciences this was an emphasis on mathematics and astronomy as providing tests of truth and, on a lower level, on medicine as a means of preserving health and beauty.

THE TRIUMPH OF GEOMETRY

From the moment of the discovery of the irrational (p. 178), Greek mathematicians turned away from numbers to the consideration of lines and areas, in which such logical difficulties did not arise. The result was the development of a *geometry* of measurement which is perhaps the chief gift of the Greeks to science. Babylonian mathematics and its successors in India and Islam remained primarily arithmetical and algebraic. The chief architects of this transformation were Hippocrates of Chios, *c.* 450 B.C., and Eudoxus, 408–355 B.C. The former was the first to teach in Athens for money and the first to use letters to denote geometrical figures. He occupied himself with the geometrical solution to the classical problems of squaring the circle and doubling the cube. Though he failed in both, he established chains of valuable propositions in the way in which Euclid was later to build his Elements. These problems, together with that of the trisection of an angle, which cannot be solved by ruler and compass, led other geometers like Hippias of Elis to construct higher curves and open a new branch of geometry.

Eudoxus was probably the greatest of Greek mathematicians. It was he who founded the theory of proportions applicable to all magnitudes and discovered the method of exhaustion or successive approximation for measuring lines and areas which after it had been extended by Archimedes was to be the basis of the infinitesimal calculus.

SPHERICAL ASTRONOMY

In the same period came the logical development of Pythagoras' world picture. Here the master was the same Eudoxus, as great an astronomer as he was a mathematician. He was able to explain the motions of the sun, moon, and planets by means of sets of concentric spheres each rotating about an axis fixed in the one outside it. The model was crude and mechanical but could serve at the same time in the form of actual metal spheres as a method of observation far more flexible than the old gnomon or dial. It is one from which all astronomical instruments to this day are derived. The theory of spheres was simple, too simple indeed to explain even the facts known long before to the Babylonians, such as the shorter length of the seasons of autumn and winter, which take up 89 days 19 hours and 89 days 1 hour respectively compared to spring and summer, which are 92 days 20 hours and 93 days 14 hours respectively. At the time these seemed minor blemishes to be removed by adding more celestial clockwork – a process which went on generating complexity until it was all swept away by Copernicus and Newton.*

GREEK MEDICINE: HIPPOCRATES

Greek medicine furnished yet another contribution to a coherent scientific world picture. It wove the two strands, one empiric, one philosophic, that have run through medicine ever since. Greek medicine, like Greek mathematics, is in unbroken continuity with the medicine of the ancient civilizations (pp. 124 f.). The Greek doctors seem to have belonged to the Asclepiadea, or clan of Asclepius, the demi-god of Medicine, one of the occupational clans or guilds. Indeed, we have still in the Hippocratic oath [2.83.332] a well-preserved relic of an adoption ceremony into the clan, carrying with it certain obligations to clan members and their families, which are still observed today. For instance, we find in it the clause:

> I will impart it by precept, by lecture and by all other manner of teaching, not only to my own sons but also to the sons of him who has taught me, and to disciples bound by covenant and oath according to the law of the physicians, but to none other.[1.71.213]

In Greece, as in the old civilizations, the doctor was somewhat of an aristocrat, dealing mainly with wealthy patrons. The treatment of ordinary people remained in the hands of the old wives and charlatans using traditional and magical remedies.

The first trend in Greek medicine is that associated with the almost legendary physician – Hippocrates of Cos. The so-called Hippocratic corpus is a mass of medical treatises, probably written over the period from 450 to 350 B.C., the tone of which is resolutely clinical. Medicine is treated as the art – *technè* – of curing patients. The most famous quotation from Hippocrates was occasioned by the need to warn physicians not to feed patients suffering from fevers:

> Life is short, and the Art long; the opportunity fleeting; experiment dangerous, and judgement difficult. Yet we must be prepared not only to do our duty ourselves, but also patient, attendants, and external circumstances must co-operate.[1.71.229]

49. A collection of Greek and Roman surgical instruments: a Roman cryptotome (left), a Greek gouge, and a curette (right). Now in the Wellcome Medical Museum.

Each case is considered on its merits, but the opinion on it is based on observations of similar cases. In this it follows the tradition of the Egyptian doctors (p. 125). Magical or religious causes or cures for disease are not mentioned, and Hippocrates goes further in an explicit renunciation of such causes. Thus in the passage on the 'sacred' disease, epilepsy, we find:

It seems to me that the disease called sacred is no more divine than any other. It has a natural cause, just as other diseases have. Men think it divine because they do not understand it. . . . In Nature all things are alike in this, that they can all be traced to preceding causes.*1.70.4

The school of Cos is, moreover, equally intolerant of the application of philosophy to medicine. In *Ancient Medicine* (the author of which may be the sophist Protagoras) we find:

All who attempt to discuss the art of healing on the basis of a postulate – heat, cold, moisture, dryness, or anything else they fancy – thus narrowing down the causes of disease and death among men to one or two postulates, are not only obviously wrong, but are especially to be blamed because they are wrong in what is an art or technique [*technè*], and one moreover which all men use at the crises of life, highly honouring the practitioners and craftsmen in this art, if they are good.[2.24.63]

Despite this denunciation, the use of philosophical postulates tended to increase in medicine and even to find its way into Hippocratic writings.

In part this arose from the beginning of anatomical and physiological studies. A follower of Pythagoras – Alcmaeon for instance – learned by dissection something of the function of nerves and dared to assert that the brain, instead of the heart, was the organ of sensation and movement. This fact, which must have been known practically to primitive hunters, was still being stoutly denied by doctors 2,000 years later. The more mystical doctrines found much more ready acceptance. Another Pythagorean, Philolaus, formulated the doctrine of the three spirits or souls of man: the vegetative spirit, which he shares with all growing things, situated in the navel; the animal spirit, shared with beasts only, which gives sensation and movement, in the heart; and the rational spirit, possessed only by man and located in the brain. These spirits were to haunt physiology and anatomy for centuries and prevent men using the evidence of their senses, till Harvey laid them to rest (pp. 437 f.).

THE DOCTRINE OF THE HUMOURS

The most persistent and damaging to the practice and theory of medicine was, however, the doctrine of the four humours, which was firstly put forward by Empedocles (pp. 174 ff.). He was a doctor as well as a philosopher and he naturally extended his cosmological ideas into his medical theory. He considered that the same four elements or 'roots of things', of which the universe is made, must be found in man and in all animate beings. For him, following probably more ancient and mythical models, man was a microcosm – a small world modelling in himself the macrocosm, the great world. The four elements of the world – fire, air, water, and earth – were matched by the four humours of the body – blood, bile, phlegm, and black bile. These are also the four sacred colours of alchemy – red, yellow, white, and black.* According to which was

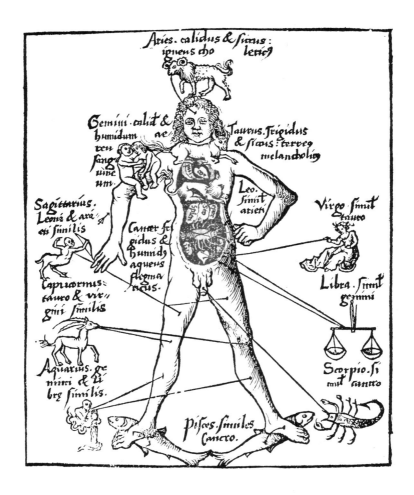

50. The doctrine of the humours of the body was also linked with the supposed influences of the celestial bodies, the signs of the zodiac acting through the planets. Each sign acted on a different part of the body, each planet on a different organ. They had power to produce disease and alter the balance of the humours, which are also mentioned on this woodcut. From Gregory Reisch, *Margarita Philosophica*, Heidelberg, 1508.

predominant, the man was sanguine, choleric, phlegmatic, or melancholic. This led to a whole system of apparently rational medicine, which was for centuries to supersede the practical art of medicine of the original Hippocratic school (pp. 278, 398 f., 564 f.). On this theory treatment was aimed at restoring the appropriate balance of the elements by controlling the two poposite pairs of qualities, hot and cold, wet and dry, which determined the elements. Fire was hot and dry, air hot and wet, water cold and wet, earth cold and dry. If a man had a fever he needed more cold, if a chill more heat.

It is easy to see now that these theories bore practically no relation to the facts of physiology, and that medical practice based on them could rarely if ever have good effect. Unfortunately, in spite of their careful clinical studies, the school of Cos were also in no position to prescribe effective treatment. They excelled in prognosis and relied on the patient, if not given violent or unsuitable treatment, getting well through the curative power of Nature. Accordingly the profession naturally preferred a doctrine in which they had a larger share in the cure, and which exalted their art into a philosophy worthy of being followed by the best people.

4.6 The Athenian Achievement

THE SOCIAL PHILOSOPHY OF ATHENS

In the second and central period of Greek thought the interest of philosophy, which still included science, shifted from the material to the ideal plane. In this it reflected the later stages of the dramatic culmination of the development of the city state in the Athenian empire in the fifth and fourth centuries B.C.[2.82] These events, because they revealed new forces at work in society and because they were so clearly and beautifully set down for future generations in the works of historians like Thucydides, remain of enormous importance to science and politics to this day. It began with the rise, for the first time in human history, of a deliberately constituted citizen democracy. That democracy remained in power for long enough to show something of its enormous creative possibilities, to which the Parthenon and the Athenian tragedies still stand witness. It fell in the end because it was based on slavery and the exploitation of foreign territory. It was unable to resist the attacks of aristocratic reaction, embodied in the far more primitive State of Sparta well supported by Persian gold.

The failure of Athenian democracy marked the turning point of classical civilization. Never again was it to approach as close to a popular control of social life and the overthrowing of the rule of the wealthy. From then on the Greek city state, for all its material successes and even its intellectual achievements, was doomed to ultimate destruction. Democracy had come near to offering real escape from the economic contradiction of the iron age city; without it the only other path was towards an increase of slavery at home and military adventures abroad. For another five centuries this was to spread Greek civilization over a large part of the world, but its inner development had come to an end.

THE PHILOSOPHERS OF REACTION

The great triad of Greek philosophers, Socrates, Plato, and Aristotle, all belong to Athens, but to the Athens of decline. They drew their enormous ability and power to influence thought from the revolutionary greatness of the first free city; the service they put it to was that of counter-revolution. Socrates, at least as Plato represented him, Plato himself, and Aristotle all showed a scorn of democracy that only partially hid their deep fear of it. Marx was too kind to the philosophers, or perhaps he was thinking of his old favourite Epicurus, when he said: 'The philosophers have hitherto only tried to understand the world, the task, however, is to change it.' The task that Plato quite consciously set himself was that of *preventing the world from changing* – at least in the direction of democracy.

SOCRATES AND LOGIC

This idealistic reaction in Greek thought was expressed in terms of the new technique of *logic* or the handling of words = logoi. Athenian politics in the democratic era gave to disputation and oratory an even greater importance than they had in most Greek cities (pp. 163 f.); they were a recognized way to fame and wealth. This gave rise to a new interest in words and their meanings. The control of people by words became more rewarding than the control of things by work. A whole new class of professional wise men – the sophists – arrived to teach this road to success to those who were willing to pay. The most famous of them, Protagoras, is remembered for the saying, 'Man is the measure of all things,' expressing the primacy of human convention over any absolute knowledge. His opponent was Socrates himself, who developed a method of argument in which, by asking a series of questions directed at his opponent's own knowledge, he could in a very short time make it clear to the audience that his opponent did not know what he was

SOCRATES TRIUMPHANS.

51. The death of Socrates, as illustrated in the frontispiece to the first English translation of Plato's *Apology* and *Phaedo*, published in London in 1675. The translator is not named.

talking about. For Socrates the chief end of man was individual goodness or virtue which was to be an automatic result of knowledge. Both the Greek word for goodness, *arete*, and the Latin, *virtus*, originally referred to combative manliness. Ares was the god of war. It took a long time to soften into the ideal of citizenship and still longer to Christian submissiveness. According to Socrates the knowledge which led to goodness was not physical knowledge or indeed anything that could be learned, it was rather a rejection of all *opinion* and reliance on inner *intuition*. In this he resembled his contemporary the Chinese philosopher Lao-tze, who was as sceptical of convention and as secure in the hold on an inner natural truth.

Socrates had his private 'daemon' who inspired him at critical moments. What his own beliefs were it is difficult to say, because he wrote nothing and nearly everything we know of him comes from Plato. Socrates was a wonderful talker and a great character and had an enormous influence on the Athens of his day, making both devoted friends and bitter enemies.

Although himself a man of the people he was not a supporter of democracy, and mingled, at least in his later years, mainly with the rich and with aristocratic young men. Some of these, like Alcibiades, turned against the city in the Spartan war, while others, like Critias and Charmides, took part in the reactionary government of the thirty tyrants formed after the defeat. These were turned out by a popular revolt in 403 B.C. and replaced by a democracy which, however, was pledged to the Spartans not to take political reprisals. It was under this government that Socrates was accused of impiety and of corruption of the youth, but the real reasons for his trial were political. His enemies apparently only wanted to exile him, but his calm and defiant defence led them to sentence him to death and make him the first and most famous martyr of philosophy. The circumstances of his life and death, more even than his own character, mark a parting of the ways in Greek thought. Henceforth philosophy will have a moral or ethical and a natural or physical branch, and for 2,000 years the first will have the greater prestige.

PLATO

Plato, as a wealthy young Athenian aristocrat, came under the influence of Socrates at a time when his political ambitions seemed thwarted for good by the return of democracy.[2.6a] He determined to devote his life to philosophy with the object of leading men to a better life by working out the principles of a perfect State. This led him on to the path of *idealism* in philosophy, and indeed he became for all time its greatest exponent. For though he was far from being the first idealist, he was able to present his views in the form of dialogues with a beauty and persuasiveness that have never been bettered in philosophical writings. Indeed their beauty of expression has hindered men in all ages from seeing the ugliness of the ideas expressed. The main political objective of Plato, expressed especially in *The Republic* and *The Laws*, is to draw up the constitution of a State where all the old privileges of the *aristocracy* – the best people – will be preserved for ever, and which at the same time can be made acceptable to the lower orders. For inspiration he turned to Sparta, where the barrack life the citizens led in common was supposed to preserve them from graft and political intrigue, and to keep down the helots,[2.89] though it had notoriously failed to do the first and ultimately the second as well. Plato divided the citizens of his Republic into four grades: the guardians; the philosophers, who ruled; the soldiers, who defended; and the people, who did all the work. The guardians held everything in common without even family life. The common people were allowed this luxury but no power at all. These class divisions were to be permanent and justified by a myth or 'noble lie' about God creating men of four kinds – gold, silver, brass, and iron.

These are the four colours yellow, white, red, and black that already appear in the humours (p. 188), and are also the *varna* of the original castes of India: Brahmins (sages); Kshatryas (warriors); Vaishnavas (cultivators); and Sudras (outcastes). Cornford, however, claims that Plato was not thinking in class terms and that each class was chosen so as to be most suitable for its duties. However, the passage he quotes hardly bears this out. In Plato's allegory,

if the rulers find a child of their own whose metal is alloyed with iron or brass, 'they must, without the least pity, assign him the station proper to his nature, and thrust him out among the craftsmen and farmers. If, on the contrary, these classes produce a child with gold or silver in his composition, they will promote him, according to his value, to be a Guardian.'[2.18.133]

This shows clearly that normally the classes were hereditary but that Plato, like the British ruling class of today, was clever enough to see that to allow a limited number of able members of the lower orders into the upper classes was the safest method of perpetuating their rule.

Through this rigid class system Plato hoped to find perfect and above all stable government. The guardians would have no responsibility to their families but only to the State, and no material cares or ambitions. They would also be subjected to an education in philosophy, mathematics, and music, which would, he thought, induce a superior benevolence. In this way he hoped to graft on to the Spartan constitution some of the remembered glories of the Athens of Pericles, where for some years the new democracy had entrusted the rule of the city to a cultured group of wealthy citizens. Plato hoped to get his political views accepted by finding a prince who was a philosopher or who could be trained to be one. His last effort was with Dionysius the younger, tyrant of Syracuse, but neither he nor his court could stand the rigours of the mathematical training required. Plato's republic has been variously judged by later generations. In the Middle Ages, compared with the arbitrary and inefficient rule of illiterate kings and nobles, it seemed a progressive ideal, especially as it was presented in such beautiful and persuasive prose. In our time, however, we see in it most unpleasant anticipations of the maintenance of the class rule of the capitalists, [2.71] which found an echo in the bogus Corporate State of the Fascists.

To support this central theme of the ideal city, and at the same time to justify the life of its philosopher guardians, Plato took over the views of Pythagoras and Parmenides (pp. 181 f.) which exalted the apprehension of absolute truths that were unchanging, logical, and mathematical. The emphasis on the discussion of words and their true meanings tended to give to words a reality independent of the things and actions to which they referred. Because there is a word for beauty, beauty itself must be real. Indeed it must be more real than any beautiful thing. This is because no beautiful thing is altogether beautiful, and so whether it is beautiful or not is a matter of opinion, whereas beauty contains nothing but itself and must exist independently of anything in this changing and imperfect material world. The same logic applies to concrete things: a stone in general must be more real than any particular stone.

PLATONIC IDEALISM

Thus grew up the fantastic world of *ideals* – images of perfection – of which the material world was but a flickering shadow on the walls of the cave in which we are imprisoned in this life.[2.68]

Plato moreover was not really concerned with providing an explanation of these appearances; what seemed all-important to him was to prove that certain abstract conceptions were absolute and eternal, independent of sense impressions and to be grasped only by the eye of the soul. These were the triad of absolute values: truth, goodness, and beauty. The first he owed to Parmenides, the second to Socrates, and the third was his own special contribution, though one drawn from the art-for-art's-sake aestheticism of the wealthy Athens of his young days. These absolute *values* are still with us. The claim that they are superior to, and beyond any knowledge derived from, the senses is used now as then to put a limit on scientific investigation and to support intuitive, mystical, and reactionary views.

Yet Plato himself argued for them on the basis of such science as was known in his day. He derived them, in fact, largely from mathematics and from astronomy, or rather astrology. The word *astrology* or reasoning (*logos*) about the stars was coined by Plato himself to replace the old astronomy, or mere ordering (*nomos*) of the stars. Later astrology got such a bad name that the old word came back. He embraced and extended the mystical views of Pythagoras on the cosmic importance of number and geometrical figures, and found in them examples of absolute truth independent of the senses. Plato does not seem to have contributed much to mathematics himself, but his influence undoubtedly gave it a prestige that drew many good minds to it later. Being, however, deliberately abstract and contemplative, it drew mathematics away from its origin in, and application to, practical experience and thus held back the development of algebra and dynamics.

ASTROLOGY

With mathematics Plato coupled astronomy, but it was a peculiar kind of astronomy, the stars as they should be rather than as they were. The old popular view was that the heavenly bodies, and particularly the sun, moon, and planets, were divine beings. That is why old-fashioned people resented as impious the claims of the Ionian philosophers that they were globes of fire wandering (*planein*) through the sky. Plato saved the situation, but at a terrible cost to science; he combined mathematics with theology by asserting, in the face of evidence already existing,[2.25.77] that the planets showed their divinity by the unchanging regularity of their perfect and circular movements, composing between them the inaudible harmony of the spheres. Thus any alteration was banished from the heavens as he would have liked to have banished it from human affairs, and the highest duty of man was to contemplate eternity and to find in it the proof of his own immortality. Plato's philosophy took back the challenge science had offered to faith. By postulating celestial perfection he stifled the ideas, already expressed by the Pythagoreans, that it was the earth itself that moved. His influence was consequently effective, together with that of his great rival and successor, Aristotle, in holding back man's knowledge of the real motion of the heavens, and with it any possibility of valid physics, for 2,000 years.

THE ACADEMY

When Plato's hopes of the philosopher prince failed, he returned to Athens, being captured and nearly sold as a slave on the way. There for forty years (387–347 B.C.) he expounded his doctrines in the groves of the hero Academus to a number of very select pupils. Over the gate was written LET NO ONE IGNORANT OF MATHEMATICS ENTER HERE.

The teaching in the Academy did not stop with Plato's death. Though it did not develop his ideas significantly it preserved them and, with the prestige of Plato and Athens behind it, the Academy endured for nearly 1,000 years till Justinian closed it in A.D. 525. It was an extension and rationalization of the mystical fellowship of Pythagoras. Both discussion between initiates and the teaching of aspirants were undertaken. Its great importance is that it is the parent of all the universities and scientific societies of our day. It was Plato himself who determined the character and tone of the institution. It was certainly academic in the modern sense. Pure knowledge, almost exclusively mathematics, astronomy, and music, was to be acquired by the reading of texts rather than the study of Nature, which was full of deceptions and irregularities. Plato's insistence on mathematics, however, did secure the presence of at least one scientific discipline in what might otherwise have been a purely literary education. Confucius, whose influence on Chinese education was to last almost as long as Plato's on that of the West, omitted mathematics. This may well have contributed to the relative backwardness of Chinese science. Ideally, in the Athenian Academy the knowledge of the true, the good, and the beautiful was sought for its own sake. Actually it was considered by the later Greeks, and the Romans after them, that it was an excellent training for a distinguished career for young men of good family.

PLATONISM

The influence of Plato, however, penetrated much further than the Academy. Progressively debased by keeping the mystical elements and neglecting the logical and mathematical ones, Platonism permeated all conformist thought in late classical times. It mixed early with Christianity and indeed formed the major intellectual support for its theology. After the closing of the Academy Plato's original works were forgotten, all except the most absurd of them, the *Timaeus*, which contains his mythical account of the formation of the world. His teaching was transmitted largely through the Neoplatonism of the even more mystical Plotinus (p. 260). The Arabs rediscovered some of his other works and translated them, but it was not till the Renaissance that they were again studied in the original and had an effect at least as great as when they were first written. It was largely owing to Plato that the early humanists were not scientific. In the sixteenth and seventeenth centuries, however, the mathematical inspiration of Plato played an important part in guiding the thoughts of Kepler, Galileo (pp. 429 f.), and through the Cambridge Platonists, also of Newton (p. 481).

ARISTOTLE

Aristotle, at first a disciple of Plato, broke from the Academy after the master's death and in 335 B.C. founded a rival school of philosophy, the Lyceum. He was born at Stagira in Thrace, but belonged to the Greek clan of Asclepiadae or physicians (p. 186). Aristotle came to occupy, for a variety of reasons, a central place in the history of science. Living as he did at the culmination of one phase in Greek political life and at the beginning of another, he was in a position to collect all the knowledge of the free Greek cities and pass it on to be applied in the empires that took them over. For most of his life he enjoyed special favours from cities and kings, and he made full use of his opportunities. His scientific output was larger and covered a wider range than that of any other single man before or since. Further, most of his work reached posterity, for it was handed on, enlarged with voluminous commentaries, by the Lyceum, whch was at first as active in inquiry as the Academy in contemplation.

Aristotle was a logician and a scientist rather than a moral philosopher. He lacked the high-minded reforming zeal of Socrates or Plato. Belonging to a later generation he realized that Plato's social ideas were out of date. Plato's philosopher prince, Dionysius the younger of Syracuse, was neither able nor willing to preserve the kind of independent aristocratic republic of which Plato dreamed. Aristotle had his own prince, no less a one than the young Alexander, whose tutor he was from 343 to 340 B.C., but he was dreaming of carving out a great Macedonian military empire rather than of ruling a Greek city state.

Aristotle was content to make the best of things as they were. He was above all the philosopher of common sense, almost of commonplace. He saw no need to change the State. All that was necessary was for people to adopt a moderate course and things could go on very well as they were. This was the celebrated doctrine of the *mean* – neither too much nor too little – which was the basis of his *ethics*.

CLASSIFICATION AND FORMAL LOGIC

The great contributions of Aristotle were to *logic*, *physics*, *biology*, and the *humanities*; in fact he founded all these subjects as formal disciplines, and even added *metaphysics* for what would not fit into them. His greatest, and at the same time his most dangerous, contribution was the idea of *classification*, which ran through the whole of his work and was the basis of his *logic*. He introduced or at least codified the way of sorting things based on resemblance and difference that we still use. The questions he asked were: What is the thing like? – genus. And how

52. A mosaic from Pompeii depicting Greek philosophers; a reconstruction, per-haps, of the Academy of Plato. Now in the Naples Museum.

does it differ from the things that are like it? – differentia. His verbal dodge, the *syllogism* – all men are mortal; Socrates is a man; therefore Socrates is mortal – is still taught as logic today, as if we could ever know the general before knowing the particular.

Aristotle was the first great encyclopedist. He tried to give some account of every aspect of Nature and human life of interest in his time. Moreover, he succeeded, where many encyclopedists failed after him, in doing so in an orderly way. He inherited the order from earlier thinkers. Aristotle took over and effectively canonized the system of four superposed elements, *fire*, *air*, *water*, and *earth*, for the sublunary sphere, and even added a fifth, the quintessence, *ether*, for the upper regions.

The earth, water, and air are peopled with living things each in its proper place with its proper form. Though each individual is subject to birth and death, generation and corruption, the form remains unchanged (pp. 902 f.). Aristotle broke definitely with the Ionian school by refusing to consider how the world had been made. The world always was as it is now because that is the reasonable way for it to be. There is no need for any creation. This was somewhat of a difficulty when Aristotelianism was taken as the philosophic basis of the Catholic Church, but it was easily overcome by introducing a sudden creation in the beginning and a sudden destruction at the end and leaving everything in the middle exactly as it was.

ARISTOTLE'S PHYSICS

The key to the understanding of the world, according to Aristotle, was *physics*. But by physics he did not mean what we mean now – the laws of movement of inanimate matter. Quite the contrary. The *physics* or nature of any being was what it tended to grow into and how it normally behaved. Indeed Aristotle's thought, because of his medical background and biological interests, interpreted the world as if everything were alive. He used *physics* in the sense that *nature* is used in the hymn:

> Let dogs delight to bark and bite,
> It is their *nature* to.

The object of scientific inquiry was to find the *nature* of everything. It had to range from explaining why all stones fall, to why some men are slaves. In every case the answer is the same, 'It is their nature to.' It is, in fact, as comprehensive an answer as to say, 'It is the will of God that they should', but it sounds more scientific. As Butler expressed it of a later philosopher, Hudibras:

> He knew what's what, and that's as high
> As metaphysic wit can fly.

In Aristotle's *Physics* and *On the Heavens* he applied this method to what we call the physical universe, where it is most inapplicable. His explanation was hardly more plausible than that of Plato, and lacked both its emotional exaltation and mathematical interest. But because it was part of the great Aristotelian logical universe, it became the main form in which Greek thought on the structure of the universe was transmitted to posterity. This was to prove particularly unfortunate for the progress of physics. Giordano Bruno had to be burnt and Galileo

53. According to Aristotle, comets were fiery phenomena in the sphere of the air and as such were believed to be followed by pestilence and disease. Later they came be thought of almost exclusively as evil omens. From Conrad Lycosthenes, *Prodigiorum ac Ostentorum Chronicon*, Basle, 1557.

condemned before doctrines which were derived from Aristotle, rather than from the Bible, could be overthrown (pp. 433 f.). The subsequent history of science is largely, in fact, the story of how Aristotle was overthrown in one field after another. Indeed Ramus was not far from the mark when he maintained in his famous thesis of 1536 'that everything Aristotle taught was false'.

FINAL CAUSES

Aristotle built his physical world in the image of an ideal social world in which subordination is the natural state.[2.24.135] In this world everything knew its place and, for the most part, kept to it. Natural motion occurred only when something was out of place and tended to return to it again – as a stone falls through air and water to rejoin its native earth, or as the sparks fly upwards to join the celestial fires. This applies only to such objects as have no natural motion of their own. It is in the nature of a bird to fly in the air, of a fish to swim in the water. That, in fact, is what birds and fishes *are for*. In this can be seen one of his leading ideas, that of *final causes*, in which organisms and even matter are endowed with a purpose to reach appropriate *ends*. Aristotle admitted other causes such as the *material cause* and the *effective cause*, which provided the material support and made things work, but he considered them inferior to final causes. This doctrine has been a curse to science because it furnishes a glib way of explaining any phenomenon by postulating an appropriate end for it, without having the bother of finding out how the phenomenon works.

MOTION AND THE VACUUM

The battle against final causes in science has been a long one, and victory is by no means yet complete. According to Aristotle natural motion is final; all other motion requires a mover, as when a horse draws a chariot, the slaves row a galley, or when the unmoved mover turns the outer sphere of the heavens. What, however, is to be said of violent motion, as when an arrow is shot from a bow? This had long been a difficult question for Greek physics and Zeno had already by a triumph of logic proved that the arrow could not move at all. Aristotle found the solution: the mover was air – 'The air is opened up before and closes in behind.'

This error led to another which was to prove as great a stumbling-block to later physics. If air is necessary for violent motion and violent motion exists in the sublunary world, the sublunary world must be full of air and a *vacuum* is impossible. The syllogism is complete, but as the minor premiss is wrong the whole argument collapses. Aristotle uses another argument against a vacuum which seems in some contradiction to the first.[2.11.69] Aristotle argues:

as air resists motion, if the air was withdrawn a body would either stay still, because there was nowhere for it to go, or if it moved it would go on moving at the same speed for ever. As this is absurd there can be no vacuum.

It is interesting to see that here he states almost word for word Newton's first law of motion, and uses its *a priori* rejection to prove the impossibility of something within a few miles of his head. But in any case a vacuum would not do; to admit it would lead straight to atomism and

54. The Aristotelian concept of the path of a projectile. Since he believed no body could undertake more than one motion at a time, the path had to be composed of two separate motions in a straight line. Not until the sixteenth century and the work of Galileo was the true parabolic path realized. From Daniele Santbech, *Problematum Astronomicorum*, Basle, 1561.

atheism. The doctrine of 'Nature abhors a vacuum' had a practical origin in the experiences of sucking up liquids, which led to the suction pump. In the end it was the limitation of the suction pump that was to lead Torricelli to the production of the vacuum (pp. 469 f.).

BIOLOGY : THE SCALE OF NATURE

The inadequacy and wrong-headedness of Aristotle's physics are partly compensated by the extent and quality of his biological observations. The qualification is not the fault of Aristotle, because the valuable contributions he made to the classification and anatomy of animals received comparatively scant attention till our own times, when it was too late for them to be of any help. In biology the idea of final causes is a much more plausible one, as it is an expression of the successful adaptation of organisms to environment – 'Oh, grandmother, what big teeth you have!' 'The better to eat you with, my dear.' The big bad wolf was a perfect Aristotelian and not too bad an ecologist. Nevertheless, even in biology final causes have had a stupefying effect. All that is demanded is a guess at the purpose of an organ or organism.

The guiding idea of Aristotelian biology is that everything in Nature is reaching up to achieve what perfection it can, and that it achieves it in different degrees. This led Aristotle to draw up a scale of Nature with minerals at the bottom, then vegetables, then more and more *perfect* animals, and finally man at the top.[2.58] Such a scale might be thought to imply evolution, but Aristotle was sure that nothing really changed in the world, and that *species* must be eternal fixed signposts to perfection or imperfection. Indeed he tended far more to see a beast as an imperfect man, and a fish as an imperfect beast, than the other way round. His immense authority, added to that of Genesis, held off the idea of evolution for more than 2,000 years. The idea of different grades of perfection was useful in another way – it justified the belief that some men are naturally masters, some naturally slaves. If these are so unnatural as not to realize it, wars to enslave them are naturally justified.

MATTER AND FORM

The concept of master and slave, of order and subordination, runs through all Aristotle's thought. He expresses it in his adaptation of Plato's ideals, the dual concept of *matter* and *form*. Matter is brute, undifferentiated; form is imposed on it by mind (*nous*). The crudest matter is capable of any form. It has all forms in it *potentially*. The form represents a purpose of perfection, which may not always be reached. In making a statue, for instance, the matter is passive and accommodating up to a point, but sometimes refractory, as when it breaks the hammer or otherwise refuses to accept the form which the sculptor wishes to impose on it. As a result of this refractoriness of matter, nothing in the sublunary world is *perfect*, each particular thing has *accidental* features, where *matter* and *chance* have thwarted rational *purpose*.

SUBSTANCE AND ESSENCE

Aristotle's *forms* are distinguishable from Plato's *ideals* because they are not *universals*, each refers to a particular animal or thing. In Aristotle's terminology the forms are substantial. This word *substance* means to Aristotle something very different from its meaning in modern science.

It is a metaphysical character by which a thing is itself and no other. To allow for some measure of change while preserving individuality, underlying each substance there is an *essence*. Thus substantially a man has two legs, but these are not part of his essence, for he may lose one or both of them without ceasing to be a man. The ideas of essence and *potentiality* are biological in character, expressing the lower and upper limits of what an individual of species can reach. In the first case it just manages to exist, in the second it is exhibiting its full powers.

The idea of potentiality opens the way to the conception of evolution of forms from the imperfect to the perfect. Perfection is always conceived following Parmenides and Plato as higher and unchangeable. Living things are sensible and corruptible, higher than them come heavenly bodies, sensible and incorruptible. Higher still is the *rational soul*, insensible and incorruptible, and highest of all is God, the most changeless of all substances and hence the most actual, the most fully realizing its potentiality.

MAN AND GOD

Thus the crown of Aristotle's work was its extension to man as a social animal, *zoon politikon*, and beyond him to God. Man contained in himself, following the doctrine of Philolaus, three souls or spirits: the vegetable soul, the animal soul, and the rational soul or *nous*. The last belonged to man alone. The *purpose* of each soul, which was its motive power, was to strive for its own *perfection*; the *vegetative* soul, for *growth*; the *animal* soul, for *movement*; and the *rational* soul, for *contemplation*. The perfection of the rational soul was to strive for something even more perfect, which could only be God, the *unmoved mover* of the whole universe, at the same time the centre and the boundary of Aristotle's metaphysics. Aspiration and love can only be upward: 'We needs must love the highest when we see it', as a slave for his master, a woman for her husband, and a man for God. Love down the scale is not called for. It was this theocentric conclusion that so endeared Aristotle to the clerical schoolmen of the Middle Ages and helped them to overlook the contradiction between his philosophy and the Bible story of creation.

Taken as a whole, Aristotle's system of philosophy is a magnificently comprehensive rationalization of the experience and attitude of a reasonably well-to-do citizen. Only a mind that combined enormous industry with unshakable complacency could have worked it out. Its genius did not lie in any of its separate parts. Except for a few personal biological investigations, none of it was original; but what was borrowed was taken from the best people. Its peculiar genius lay in its comprehensiveness,

55. Aristotle's best personal scientific work lay in biology: in particular he made careful studies of some marine creatures as well as of bees and their diseases. Bee-keeping was of considerable importance in antiquity since it was from honey that sweetening was obtained. Virgil treats of bees in the *Georgics*, and this copperplate was used to illustrate a seventeenth-century edition.

its orderliness, and in the unity which was given to the whole system by his logic.

To achieve the comprehension Aristotle had made another innovation of enormous promise for the future. Instead of doing all the work himself or merely discussing it with his colleagues, as was the practice of the Academy, he *organized research*. In the Lyceum, which was probably subsidized by Alexander, Aristotle's young men collected information on nearly everything, from social and natural forms of literature to the constitutions of cities, from animals and plants to stones. What is left of the results today is the most valuable and systematic knowledge of Greek life and thought. Even more valuable is the practice of such investigations. Just as the Academy is the original of the university, the Lyceum is that of the research institute.

THE INFLUENCE OF ARISTOTLE

As will be shown in the next section (4.7) the following out of Aristotle's method of research was very soon to undermine or refute most of his own conclusions, including the central one of final causes. Indeed his views on many topics were out of date before he put them forward. His enormous influence on Arabic and medieval thought, however, came in spite of, or perhaps because of, those limitations. The finer developments of Greek science were either completely lost or, like the work of Archimedes, not appreciated till the Renaissance. They could not be understood except by highly trained and sophisticated readers, not easily to be found in the Dark Ages. Aristotle's works, however, though tough going, did not require, or did not seem to require, anything but common sense to understand them. Like Hitler, Aristotle never told anyone anything they did not already believe. No experiments or apparatus were needed to check his observations, no troublesome mathematics to derive results from them, no mystical intuition to understand any inner significance. Plato, it is true, appealed more to the imagination and had more moral fervour; but Aristotle explained that the world as they knew it was just the world as they knew it. Like M. Jourdain in Molière's *Le Bourgeois Gentilhomme* they had all been philosophers without realizing it. As long as the world remained the same Aristotle would do, but, as we shall see, the world did not remain the same.*

Taken together, the three great philosophers of the decline of Athens mark a definite arrest of the movement of ideas which had begun with the Ionian philosophers. Because the social order could no longer advance, the idea that Nature itself was changing and developing was repudiated. Philosophy ceased to be progressive and, as part of the same reaction, ceased to be materialist. Idealism in the mystical form of Socrates and Plato, or in the conformist scheme of Aristotle, took its place. Philosophy taught the acceptance of life as it was, and had nothing to offer to those who found it intolerable but that their sufferings were inevitable and were part of the great order of Nature. Such philosophy was well on the way to becoming a religion, but a religion for the benefit of the upper classes alone.

4.7 Alexander's Empire

HELLENISTIC SCIENCE

The arrest in general philosophic ideas did not, however, mean the end of practical science; indeed it was to prove a great stimulus to it. It remains true that no great comprehensive attack on the problems of Nature

56. Alexander's conquest of the Near East had ramifications as far as India and spread Hellenistic culture widely abroad. The Macedonian was a military leader of the greatest ability, but his task of conquest was assisted by internal struggles in many of the lands he subjugated. This mosaic from Pompeii, now in the Museo Nazionale, Naples, depicts a battle between Alexander and Darius.

and society was made between the time of Aristotle and that of Bacon and Descartes, for neither the medieval schoolmen nor the Arabs were, or would even claim to be, his master. Nevertheless most of the detailed achievements of Greek mathematics, astronomy, mechanics, and physiology come from the next age, that of Alexandrian or Hellenistic science. Only original thinking was lacking. The reason cannot be an intrinsic one, the later Greeks were at least as clever as the earlier ones. It must be looked for in the social field, in the conditions that discourage general creativity but encourage the working out of limited fields and the development of practical applications.

The great political and economic change that followed within a century of the fall of Athens was the forcible unification of the independent rival city states by new large land empires which, however, drew their culture largely from the same source. How much this change was overdue was shown by the immediate successes of Philip of Macedon and Alexander. The cities were far too weakened by internal class struggles and divided by mutual jealousies to put up effective resistance. The new type of well-drilled and well-equipped mercenary armies could go where they liked. The levies of the old Persian empire, largely untrained peasants led by hereditary nobles, were no match for them, however great their numbers.

In every other way the Greek type of civilization which the Macedonians had simply taken over showed itself superior to those of the older civilizations it overran. In technique, in organizing ability, in knowledge, in art, the Greek way of doing things imposed itself wherever it appeared. Greek merchants and administrators followed the armies, and cities of a Greek type, though often with only a minority of Greeks, were set up, from the first and most famous Alexandria in Egypt to the farthest Alexandria Eschata (Kojand) in Afghanistan. Nor did Greek influence stop there; it spread far and wide beyond the boundaries of the Alexandrian empire. In the Far East its effect was diluted by distance, but the first Indian empire, that of the Buddhist Asoka, was a direct result of Alexander's raid; and something of Greek art, philosophy, and science spread with Buddhism as far as China. At about the same time an analogous, but quite independent movement was taking place there. In 221 B.C. the ruler of the semi-barbarian Chin State created by force of arms the first Chinese Empire of iron age type, and called himself by the name of the legendary first emperor Hwang Ti. Though his dynasty was not to last, the unity of the Empire was never subsequently lost for long at a time. All through classical times the highly civilized Han Empire bordered those of Persia and India.

The influences of Hellenism on the West were much greater, because there was less indigenous culture to be replaced. The Latin clansmen were rapidly Hellenized, influenced partly by the city culture of the Etruscans,[2.64] themselves coming from Asia, and partly by that of the Greek colonists of the coastal cities. One city, Rome, after expelling its Etruscan kings, proved more powerful than all the rest, and after a stormy internal political history emerged as the plutocratic republic, which was later, as the Roman Empire, to dominate the whole area.

THE HELLENISTIC CITY, AND THE MACEDONIAN EMPIRES

The Hellenistic cities differed in many ways from the Greek cities on which they were modelled. First, to the class distinctions that already operated there was added a race or culture distinction between the Greek-speaking official and trading classes, and the natives. These natives in the south and east, though politically suppressed, knew they had a culture of their own far older and by no means inferior to that of the Greeks. Though this division softened with time it lasted to the very end of the classical era, where the old cultures reasserted themselves in a new religious form (p. 262). Secondly, the cities were not independent, but formed part of the shifting empires of the Ptolemies of Egypt, the Antiochids of Syria, and the various dynasties of Asia Minor and Greece. This was a return, though only a partial one, to the state of the old empires with a divine king, a court, and an army, originally Macedonian but later filled with every kind of local levy or mercenary. The citizens might suffer from tyrannical or, even worse, from weak kings, but they could do very little about it. The real decisions were made at court or on the battlefield. They therefore concentrated on making money and enjoying life, while the poor, the natives, and the slaves endured the situation as best they could. The result was the splitting of society to a degree that had never been reached in human history. The citizens had the opportunities to develop a very select and superior culture, but it was doomed to sterility from the start.

THE PHILOSOPHIES OF ACCEPTANCE

The spread of Hellenism had indeed taken place at the expense of its internal cultural development. In art, drama, literature, and politics, the late Greek achievements, particularly those of Athens, were, so to speak, frozen. Good models were copied in the slightly exaggerated and sentimental Hellenistic style; commentary and criticism flourished, but nothing really great and new was produced.

In philosophy there were no real successors to the schools of Democritus, Plato, or Aristotle. Indeed philosophy, which had already parted from science, was from the time of Alexander also parted from political life and became almost exclusively moral. The citizen might now enrich himself, but he no longer shared in ruling the State except by favour of the court. Philosophy was now concerned with reconciling politically impotent man to the uncertainties of life in an economically insecure and war-ridden world. The *Cynics* and *Sceptics* shrugged their shoulders. The *Stoics* put up a fine show of superior indifference based on the belief in the intrinsic value of virtue, and in a world ruled by unalterable fate which the stars determined. The *Epicureans* urged men to make the best of it, to practise virtue as the surest way to pleasure, and not to worry about the gods who lived far above this world of whirring atoms.[2.4] The philosophy of the ancient world was to peter out in the mysticisms of the Gnostics and the Neoplatonists, and the last echo of its old voice was to be the *Consolation* of Boëthius, at the end of one era but the beginning of another. Between them the philosophers represented what might more properly be called the *religion* of the cultivated upper classes. They were indeed providing the intellectual language in which the cruder, but far more vital, religions of the lower classes were to express themselves as soon as they came to power.

HELLENISTIC SCIENCE

The one exception to the general intellectual decay was for a few centuries the development of natural science. There was indeed in certain directions, particularly mathematical, mechanical, and astronomical, a notable new outburst of creative thought. This arose largely on account of the economic and technical consequences of the conquests of Alexander. By throwing open to Greek trade a world far wider than it ever had known, it created a new market which for a time relieved the chronic crisis of the Greek city state: the under-consumption due to the wretched conditions of the poor and the slaves. The export market for manufactured goods was still a class-restricted one: only goods for wealthy households were produced – chased silver, moulded pottery, blown glass, dyed cloth, papyrus, elaborately patterned textiles – but it was large enough for these goods to be made in quantity. This led to the rise of manufacturing towns employing, for the most part, wage workers kept down by slave competition. At the same time the larger areas under a single government favoured a limited sea trade in the necessities, particularly corn, to feed this non-agricultural population. This led in turn to technical improvements, not only in manufacture but also in agriculture, in which slave gangs were being used on a larger scale. Such improvements were the concern of rulers and hence of their scientific advisers. Another and even more pressing need for new techniques lay in the almost permanent state of war between empires, for which ever more complex engines were always in demand. The Macedonian rulers of the Hellenistic States were, unlike the Romans who were to supplant them, brought up in the aura of the prestige of Greek learning; they not only permitted, but encouraged it in all its branches. It was Greek science rather than literature or philosophy that was the main beneficiary.*

THE MUSEUM OF ALEXANDRIA

Indeed the great contribution of Greek science to the science of later times was for the most part derived from the work in the early Hellenistic or the Alexandrian period (330–200 B.C.) and largely at Alexandria itself, the most important Greek city of the new empire of the successors of Alexander – the Ptolemies. Greek science was brought into direct contact with the problems as well as the technique and science of the old Asian cultures, not only those of Egypt and Mesopotamia but also to a certain extent those of India. And now, for the first time in human history, there was a deliberate and conscious attempt to organize and subsidize science. The Museum at Alexandria was the first State-supported research institute, and although its artistic, literary, and even philosophic production was negligible apart from its preservation of ancient texts, it contributed more to science than any single institution had done before and possibly has done since. The scientific work of the Museum, taken together with that of its ex-members and correspondents in the rest of the classical world such as Archimedes, was far more specialized than any other had ever before been or was to be for another 2,000 years. It reflected the isolation of the Greek citizen to an even greater degree. The scientific world was now large enough to provide a small, appreciative, and understanding *élite* for works of astronomy and mathematics so specialized that even the average educated citizen could not read them, and at which the lower orders looked with awe mixed with suspicion. This enabled the scientists to venture into complex and refined arguments, and by mutual criticism to make enormous and rapid advances. At the same time these advances were very insecure. The whole scientific effort depended on the patronage of an enlightened State. When that went the edifice of learning largely collapsed and, because it had no living roots outside the big cities, was largely forgotten, though it left a few vitally important writings to be brought to light again in the Renaissance.

The main trends of work in the early days of Alexandrian science followed those of Aristotle and his school. The Museum might indeed be considered as the Egyptian branch of the Lyceum, which, as it was better endowed, in a few years came to overshadow the early foundation. Strato, *c.* 270 B.C., the most generally competent of the Hellenistic scientists, taught both at Alexandria and Athens and was the last important head of the Lyceum.

The scope of research at both institutions did not, however, embrace the whole of Aristotle's vast programme. His own biological and sociological interests were not developed further except by his immediate successor, Theophrastus, who did for botany what Aristotle had done for zoology and who began a descriptive mineralogy which, though crude, was not substantially improved for 2,000 years. It was especially physics in its *astronomical, optical,* and *mechanical* branches that was intensively studied. Instead of Aristotle's preoccupation with *logic* there was a rapid development of *mathematics* along Platonic lines. This was primarily concerned with the inherent beauty of ideal forms and the need to impress them on the merely observable world. Nevertheless it could be, and was, used on a lower plane to provide more exact astronomical descriptions and to reduce mechanics, pneumatics, and hydrostatics to exact sciences.

With ideal conditions for work, improved instruments, and scope for experiments, the cruder intuitions of Plato and Aristotle were soon left behind. Teleology, the doctrine of natural places and final causes, was abandoned, so was the Aristotelian theory of motion which made a vacuum impossible. Much of the atomic theory of Democritus, which the Athenian philosophers had so sternly expelled, was readmitted. To a great extent the first stage of the destruction of the philosophy which the Middle Ages believed to be that of the Ancients had been accomplished by the beginning of the third century B.C. Boyle would have found himself in complete agreement with the views of Strato. But he was never to learn them. Except in mathematics the advanced thought of Hellenistic times was largely lost. The reason for this has already been touched on. It was the effective isolation – social and ideological – of the scientists of Alexandria, Athens, and Syracuse. They were no longer philosophers. Strato, according to Cicero, 'abandoned ethics, which is the most necessary part of philosophy, and devoted himself to the investigation of Nature'. They therefore drifted out of the main current of interest, which in those times of crises and decadence turned inward on the inner world of the individual. Their advanced views were not propagated and, except in astronomy, where they were still needed for the more limited tasks of the time, particularly astrology, they were forgotten while the more common sense and unscientific views of Plato and Aristotle were carefully preserved.

HELLENISTIC MATHEMATICS: EUCLID

The mathematical and physical sciences were pursued in the Hellenistic world with two ends in view, the academic and the practical. The academic, which was of course the higher, was centred on mathematics and led to an extension and systematization of one branch – geometry. Numerical calculations were considered definitely inferior and were disguised as geometry when needed. But here solid and admirable results were obtained. Archimedes applied and improved these methods of Eudoxus (p. 185) to determine the value of π to five places – the practical squaring of the circle – and to find the formulae for the volumes and surfaces of spheres, cylinders, and more complex bodies. This was the effective beginning of the infinitesimal calculus which was to revolutionize physics in the hands of Newton. There was a great study of higher curves for the purpose of solving the classical and useless problems of trisecting an angle and doubling a cube. Of far greater ultimate significance was the elaboration by Apollonius of Perga, *c.* 220 B.C., of the studies of the conic sections – ellipse, parabola, and hyperbola – discovered by Menaechmos in *c.* 350 B.C. His work was so complete that it could be taken up unchanged by Kepler and Newton nearly 2,000 years later for deriving the properties of planetary orbits.

Even more important than their separate achievements was the systematization of mathematics achieved in Hellenistic times. Logical linking of theorems was known before (p. 185) – indeed Aristotle's logic is a copy in words of the geometrical procedure of proof. It was, however not until Euclid (*c.* 300 B.C.) that a large part of mathematical knowledge was built together in one single edifice of *deduction* from *axioms*. The value of this for mathematics was considerable, as shown by the fact that Euclid is still in one form or another the basis of geometrical teaching. Its value in physical science is more doubtful, emphasizing as it did the superiority of *proof* over *discovery* and of *deductive* logic based on self-evident principles over *inductive* logic based on observations and experiments. The success of geometry held back the development of algebra, as did the very primitive Greek number notation. A partial exception is the work of Diophantus, *c.* A.D. 250, on equations. This

57. The Greek flair for geometry came to a climax with the systematization of this knowledge by Euclid, *c.* 300 B.C. This page with diagrams appears in a very early printed edition of Euclid, the *Opus Elementorum* published by Erhard Ratdolt, Venice, 1482. This book is one of the most handsome productions of the 'incunabula' period of printed books and is the first to have printing in gold.

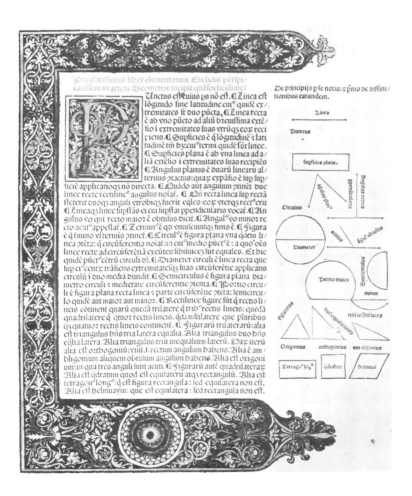

work, which comes late, shows internal evidence of the influence of contemporary Babylonian–Chaldean mathematics.*

HELLENISTIC ASTRONOMY : HIPPARCHUS AND PTOLEMY

The study of astronomy lay midway between the theoretical and practical. According to Plato it was the study of an ideal world in the sky, suited to the dignity of the gods that lived there. Any deviations which could be observed in the real sky were to be ignored or explained away. On the other hand the implied importance of the skies required that the position of the stars, and particularly of the planets, should be accurately known, and known in advance, if there was to be any hope of dodging the predictions of astrology. As a result of these two tendencies, Hellenistic astronomy – the only part of Greek science to come down to us without a break – was largely engaged in trying to make ever more complicated schemes fit the observations without violating the canons of simplicity and beauty. This pursuit encouraged the development both of mathematics and physical observation. It may be said that astronomy, almost up to our own time, was the grindstone on which all the tools of science were sharpened.

The mathematical basis of astronomy was the spheres of Eudoxus, but for actual working out it was easier to consider planetary motion in the flat and to save the appearances by introducing 'wheels within wheels'. This was done by the greatest observational astronomer of antiquity, Hipparchus (190–120 B.C.), who invented most of the instruments used for the next 2,000 years and compiled the first star catalogue. His planetary system, though more accurate, was far more complicated than that of Eudoxus and removed its last shred of mechanical plausibility. In the form in which it was presented by Ptolemy (A.D. 90–168) 200 years later it was to be the standard astronomy till the Renaissance. It was accepted because it removed all the difficulties from earth to heaven, where, after all, there is no reason to expect that vulgar mechanics would hold. Further, as it was made to measure – epicycles being added as required – it gave tolerably accurate predictions.[2.3; 2.47; 2.48]

The alternative tradition, that it was the earth that turned, put forward by Ecphantus in the fourth, or perhaps by Hicetas in the fifth century B.C., had never been lost. It was powerfully supported by Heraclides of Pontus (c. 370 B.C.), who adopted the system of a revolving earth still in the centre of the universe round which the moon and sun turned, but with the planets turning round the sun and not the earth. This system, which completely describes what is observed, was later to be that of Tycho Brahe (p. 421). The final logical step was taken by Aristarchus of

Samos (310–230 B.C.), who dared to put the sun and not the earth in the centre of the universe. This system, however, despite the eminence of its propounder, won scant acceptance largely because it was thought to be impious, philosophically absurd, and violated everyday experience. It remained, however, a persistent heresy transmitted by the Arabs, revived by Copernicus, and justified dynamically by Galileo, Kepler, and Newton (pp. 406 f., 420 f., 475 f.).

SCIENTIFIC GEOGRAPHY

The development of astronomy made a metrical and scientific geography possible for the first time. The problem of constructing a *map* is one of relating the astronomic positions on a sphere, the imaginary parallels of latitude and meridians (midday lines) with the positions of towns, rivers, and coasts, as reported by travellers and officials. This is equivalent to measuring the *size of the earth*, which was first achieved by Eratosthenes of Cyrene (275–194 B.C.), a director of the Museum. The value he found for the circumference – 24,700 miles – is only 250 miles wrong, and was not improved on till the eighteenth century. The conquests of Alexander had greatly enlarged the boundaries of the world

58. Ptolemy, the last great astronomer of antiquity, was also a leading geographer and his map of the world is to be found repeated even as late as the mid sixteenth century. From Peter Apian, *Cosmographia*, 1553.

known to the Greeks, but there they stopped – there was no economic drive to further exploration east or west, apart from a few lone voyagers like Pytheas of Marseille (*c.* 330 B.C.), until the time of the Renaissance. The lack of interest in ocean voyages made it unnecessary to develop an accurate navigational astronomy, for coastal voyages could well enough be made with a very elementary knowledge of the stars.

Optics was also a minor appendage of astronomy. The Ancients never achieved a lens – their glass was too full of flaws and crystal was too rare. Their catoptrics – the study of reflections in mirrors – was developed to the extent of arranging illusions and burning mirrors, but had no serious use. On the other hand their dioptrics – the measurement of angle by sights – was used in accurate surveying. In spite of this they never seem to have realized true perspectives, which had to wait till the Renaissance.

HELLENISTIC MECHANICS: ARCHIMEDES

It was in mechanics that the Hellenistic age furnished its greatest contribution to physical science. The first impetus probably came from the technical side. Greek workmanship, particularly in metals, had reached a high level before Alexander. Transplanted to countries such as Egypt and Syria, with far greater resources at their command, it could be used to effect radical improvements in all machinery, especially those of irrigation, weight shifting, shipbuilding, and military engines. We know that a great crop of apparently new devices appeared around the third century B.C., but their origin is still obscure. They may well have come from the discovery by invaders of traditionally developed machinery of local craftsmen, afterwards written up and further developed by literate Greek technicians. The mutual stimulation of accurate workmanship and precise calculation was to be observed again in the Renaissance. The compound pulley and the windlass may have come from sailing-ships, and gearing from irrigation works; but the screw seems a somewhat sophisticated invention. Some mathematicians may have had a hand in it. On the demands of their royal patrons, philosophers were by then prepared to debase themselves by considering the mathematical design of machinery. Certainly all the legends of Archimedes' war machines must have some foundation, though Plutarch says of him, 'He looked upon the work of an engineer and everything that ministers to the needs of life as ignoble and vulgar.'[2.70] Archimedes (287–212 B.C.) was one of the greatest figures in Greek mathematics and mechanics, and the last of the really original Greek scientists.[2.9] He was a relation of Hiero II, the last tyrant of Syracuse, and took a large part in the defence of that city against the Romans. He was killed, while working out a problem, by a

59. Archimedes was killed at the siege of Syracuse in 212 B.C., aged seventy-five. Among his many activities were his designs of war machinery for defending Syracuse against the Romans, and here is a representation of the burning glass with which he was supposed to have concentrated the Sun's rays in order to set fire to the attacking fleet – probably an apocryphal story. From Marius Bettinus, *Apiaria*, 1642.

Roman soldier who either did not know or did not care what he was doing. Though he was very much in the tradition of *pure* Greek science, we know from the chance discovery of his work on *method* that he actually used mechanical models *to arrive at* mathematical results, though afterwards he discarded them *in the proof*. For the most part his work was not followed up in classical times. It was only fully appreciated in the Renaissance. The first edition of Archimedes' works appeared in 1543, the same year as the *De Revolutionibus* of Copernicus and the *Fabrica* of Vesalius, and had an effect comparable with them (p. 430).

STATICS AND HYDROSTATICS

In his *elements of mechanics* Archimedes gave a full and quantitative account of the working of the simple machines and laid the foundations of the science of *statics*, a characteristically Greek analysis of the conditions under which forces would exactly balance. He was also the founder of *hydrostatics*, the laws of floating bodies, which was to have two

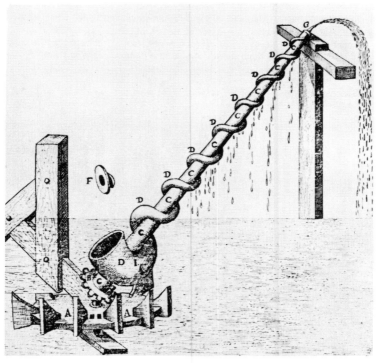

60. If Archimedes' defence of Syracuse by burning is apocryphal, his spiral screw for raising water was no chimera and is still in use in some under-developed countries. Here its principles are extolled in a book by Robert Fludd (1574–1637), *Utriusque Cosmi . . . Historia*, Oppenheim, 1617–19.

important uses. One was for the determination of the densities of bodies by weighing them in water; this, because it could be used for the testing of precious metals, was taken up at once and never lost. The other, the estimation of the burden of a ship, was well enough known by tradition to shipbuilders and was not calculated till the late seventeenth century (p. 455).

PNEUMATICS

One radically new branch of mechanics was pneumatics – the study and use of air movements. Here Ctesibius (*c.* 250 B.C.) and Hero (*c.* A.D. 100) provided many ingenious tricks working by compressed air, mostly for use in temples. Hero even constructed a rudimentary steam-engine working on a jet reaction principle. A more practical development was that of pumps. In this the technical proficiency of the metal-workers produced double-acting force-pumps as good as anything that existed before the present century and cheap enough to be used even in remote

61. Hero of Alexandria (first century A.D.) was a famous mechanician. Among his many inventions was the aeolipyle, here shown in a reconstruction drawn for a book by Dionysius Lardner, published in 1856, in which the principles of steam as a motive power are discussed. No heating of the spherical container which is filled with water is shown here, and it is doubtful if the device was more than a curiosity.

Britain. Another pneumatical device was the water-driven wind organ with stops, operated by keys just as our own organs and pianos are.

The mechanical knowledge and attainments of the Hellenistic period were in themselves quite sufficient to have produced the major mechanisms that gave rise to the Industrial Revolution – multiple drive textile machinery and the steam-engine – but they stopped short of this point. It is true they lacked the prime material of that period – cheap cast iron – but they possessed all the means to make it, power-driven bellows were well within their scope. The decisive reason was the lack of motive. The market for large-scale manufactured goods did not exist. The rich could afford hand-made goods, the poor and the slaves could not afford to buy anything they could do without.*

THE DAWN OF SCIENTIFIC CHEMISTRY

The mathematical–mechanical character of the science of the Greeks, together with their unwillingness to concern themselves with anything that would dirty their hands, prevented them from making any serious progress in chemistry, though the beginnings of alchemy and the key chemical process of distillation may date back to early Alexandrian times. Whether alchemy and with it scientific chemistry originated in Alexandria is still an open question. The first reliable writings, such as those of Zosymus of Panopolis and Mary the Jewess, come very late in the fourth and fifth centuries A.D. Any theory they had may have been affected by the influence of Chinese alchemy (p. 280). The technical achievements of Hellenistic chemistry, on which the whole of modern chemistry rests, were due to improvements in glass blowing, needed for the still (ambix) (p. 279), and in the preparation of pure materials.[2,43]

NATURAL HISTORY

Little need be said about the achievements of the Hellenistic scientists, other than doctors, outside the field of the physical sciences. The impetus given by Aristotle to a complete study of all aspects of the universe did not last more than a generation. Only a few significant advances were made in the study of animals and plants though a beginning was made on books on practical agriculture.

HELLENISTIC MEDICINE: GALEN

It was in medicine, even more than in astronomy, that the social conditions of Hellenistic and Roman times favoured a continuity of tradition and even a limited advance. The rulers and the wealthy citizens could not do without doctors. Indeed the increasingly unhealthy life they led made them more and more dependent on them. The Museum encouraged much research in anatomy and physiology.

Herophilus of Chalcedon (*fl.* 300 B.C.) was a great anatomist and physiologist basing himself on observation and experiment. He was the first to understand the working of nerves and the clinical use of the pulse, and distinguished the functions of the sensory and motor nerves. Erasistratus (280 B.C.) went further and noted the significance of the convolutions of the human brain. Although most of the finest work of the early Alexandrian period has been lost in the original, the essence of it was passed on in the tradition and was incorporated in the vast production of the last of the great classical doctors, Galen (A.D. 130–200). He was born in Pergamum in Asia Minor, but after training there and at Alexandria ended by taking a very lucrative practice in Rome. He in turn became the fount of Arabic and medieval medicine and anatomical knowledge, and acquired a reverence and authority as great in his field as that of Aristotle. The doctors of later times, impressed by his range of knowledge and experimental skill, hesitated to pit their own observations against his. Indeed, the Galenic system was a skilful blend of older philosophic ideas, like the doctrine of the three spirits or souls (p. 188), with acute but often delusive anatomical observations, largely because he was limited to dissecting animals. Galenical physiology, with its ebbing and flowing of spirits and blood in arteries and nerves, with the heart as the origin of heat and the lungs as cooling fans, still indeed lives in popular language. It was as much the basis of human belief about the little world of man – the microcosm – for over 1,000 years as Aristotle's cosmology was about the great world of the heavens. It was not until the Renaissance had got behind it a comparable mass of observations, and was furnished with a far better mechanical philosophy, that Galen's views could be superseded. How thoroughly that has been achieved is shown by the fact that the first full English translation of Galen was published only in 1952.[2.31]

4.8 Rome and the Decadence of Classical Science

By the middle of the second century B.C. the Hellenistic empires were collapsing in anarchy and under the weight of the more vigorous power of Rome. There was nothing mysterious about its success in achieving power over the Mediterranean world. Whichever native city managed to establish itself as dominant in Italy would have an enormous advantage both over the Greek or Phoenician city states and over the Asiatic Hellenistic empires, all of which had suffered from centuries of wasteful exploitation which had left them politically and economically weakened. Italy was still, in the third century B.C., a farming country with a good climate and plenty of timber, just in the first flush of expansion, with a growing, healthy population. Its slow early growth had left Rome far nearer the clan organization of society than the cities of the older civilizations. The Roman republic could count in its wars on popular support, which the others could never do. Arming themselves repeatedly with the techniques of their more advanced enemies, the Romans could be beaten in battle but they could not be conquered. Rome's only serious rival was the commercial republic of Carthage, which could match it in wealth but not in manpower.

Internally Rome had experienced essentially the same class struggle that had racked the Greek cities, but in an even more naked form, expressed in the rivalry of patricians and plebeians for control of the State. In the first century B.C., this culminated in bitter civil wars which paved the way to military dictatorship and later to empire. Indeed, the acquisition of the Empire was one means by which the rich could buy off the poor with a small portion of the loot of provinces. Another was the policy of extending Roman citizenship first to Italian and then to other provincials, thus turning what was originally a city state into a territorial State, dominated by slave-owners and wealthy merchants. Piece by piece the States of the eastern and western Mediterranean fell into Roman hands and at the same time they opened up the barbarous hinterlands of Gaul, Britain, western Germany, and Austria. The result was the formation of a great new empire, occupying the whole Mediterranean area, but sharing the Hellenistic kingdoms with a newly liberated Persia.

The cement of the Empire was the army by which it had been won, and by which, with decreasing success after the time of Augustus, it was defended against barbarians. The emperor, as commander-in-chief, usually managed to impose and collect enough taxes to keep the soldiers from mutinying and choosing another emperor. The Empire was effectively a loose federation of cities managing themselves and profiting for their mutual trade from the internal Pax Romana. The best land of the countryside was farmed by slave gangs from the villas of the wealthy. The poorer areas – the *pagi* or rustic communes – were left to the natives – the *pagans* – largely following their own tribal customs (later to become the peasants of the Middle Ages and to give their name to the country or

62. A crane operated by a treadmill with human labour. Roman, third century B.C.
Original low-relief carving now in the Lateran Museum.

63. A Roman villa at Pompeii, a typical home of the wealthy citizen.

pays) or to newly settled *coloni* and freed slaves from the *villas* who gradually became serfs – *villani*, villeins, or villains.

The spread of the Roman Empire had a very different effect on culture from that of Alexander's conquests. By the time the Romans came on the scene the impetus of Greek civilization had already passed. In science and art it was already decadent. In another sense the Romans came on Greek civilization too late; their own economic system, based on wealthy patricians and their clients, was far too set to make effective use of science.

Besides, the Roman upper class, and while the Empire was being built they were the only Romans that counted, though they adopted the trappings of Greek civilization, despised it. Neither they nor the new provincials of the West added anything significant to it. The best that they could do was to pick up some of the general ideas of Greek philosophy and use them to support their own form of class rule. The elder Cato, a country diehard of the second century B.C., hated Greek science and made no bones of it. According to him Greek doctors came over to poison the Romans, and philosophers to debauch them. Cicero, a rising lawyer a century later, took a much more enlightened view. He found much to praise in the philosophy of Plato and Aristotle, which justified the rule of the best people, but suspected that the Epicureanism which his countryman, Lucretius, was introducing would shake the people's faith in the gods and hence in established order. However, the philosophy most in vogue, especially in the days of the Empire, was Stoicism. Though it had started as a philosophy of resistance, rather like early Existentialism, Stoicism's emphasis on virtue for its own sake gave the Roman administrators, and even an occasional emperor like Marcus Aurelius, a sense of sacrificing themselves, without thought of reward, for the public good. Seneca, the most distinguished of the Roman Stoics and the tutor of the artistic emperor, Nero, saw nothing odd in accumulating a large fortune – no doubt as a sacred trust.

It is customary to blame the practical spirit of the Romans for the sharp decay of science that set in about the time of the first Roman emperors. It is much more probable that the causes were deeper: they lay in the general crisis of classical society, which flowed from the accumulation of power in the hands of a few rich men (whether they were at Alexandria or Rome did not much matter), and also in the general brutalization of a population of slaves and of what we may call, from more recent analogies, 'poor whites'. Their impoverishment lowered the demand for commodities, which depressed still further the condition of merchants and craftsmen. This was an atmosphere in which there was no incentive for science, and in which the science that still existed carried on from inertia and very soon lost its essential quality of inquiring into Nature and doing new things.

PUBLIC WORKS AND TRADE

The application of existing knowledge could, however, for several centuries be made more extensively and on a larger scale than ever before. Not only could gigantic public works such as roads, harbours, aqueducts, baths, and theatres be constructed, but unrestricted trade could

64. Roman civil engineering as depicted by Giovanni Piranesi (1720–78) in his beautiful engraving of the aqueduct of Nero (37–68) which carried part of Rome's water supply.

flourish and products from all parts of the Empire could be freely interchanged. This led, for commodities like pottery, to what was practically factory production of standardized articles. However, with abundant slave labour and a market still restricted to the well-to-do classes, the master manufacturers had no incentive to take the next step of introducing machinery, and the conditions for developing an industrial revolution never arose.

ARCHITECTURE

The two characteristic contributions of Roman technology were to architecture and agriculture. The building of aqueducts, amphitheatres, and large basilicas called for the development of the arch and the arched vault – made possible by the lavish use of burnt brick and of a concrete made from lime and volcanic ash. In spite of its massive impressiveness, Roman architecture shows far less sense of the exploitation of the possibilities of arch and vault than does the medieval Gothic. It was only in the very last stages, and that in Constantinople, that the really ingenious construction of the light pendentive supported dome was evolved from Persian models.

AGRICULTURE

Agriculture could hardly become a science until far more was known of biology than could possibly be known to the Ancients. Indeed, it is hardly a science yet. The agricultural writings of the Romans, of which the best known are the *Georgics* of the poet Virgil, are necessarily limited to recordings of peasant practice together with some grim reminders of estate management based on slave labour. They are none the less interesting in showing how, particularly in fruit and vegetable gardening, most of the techniques of today were well known and practised. On the other hand, lack of suitable horse harness and ploughs set a limit to the kind of land that could be cultivated.

65. Roman agriculture epitomized in ploughing, harrowing and other activities, from a seventeenth-century edition of Virgil's *Georgics*.

ADMINISTRATION AND LAW

The great positive contribution of the Romans to civilization that is found in every history book is their creation of a system of law. Now Roman law is anything but a scientific attempt at securing fair dealing between man and man: it is frankly concerned with preserving the property of those fortunate enough to have acquired it. It contains, as Vico first saw, the relics of three superimposed layers of cultural history. First, there is the old tribal custom, evolving from its matriarchal to a most severe patriarchal stage under the influence of the monopolization of movable property in cattle (*pecunia*). This is the celebrated Roman *family* system in which the paterfamilias despotically rules his wife, children, and *famuli* or slaves. Next comes the imprint of city and merchant law, the result of the long economic and political struggles of the Republic, with its emphasis on cash and recovery of debt. Last is the effect of the Imperial administration with the recognition of the *prerogative* of the prince. In its final codified form, at the very end of the Empire under Justinian in the sixth century, it shows the influence of the severe Stoic philosophy which, like Confucianism in China, had become the second nature of the Roman officials. There is much social history to be learned from Roman law, but it contributed to science only the concept of a universal *law of nature*.[2.17] Inapplicable essentially to the totally different economy of the feudal period, it was revived, with all the aura of the greatness of the Empire, in the Renaissance as the basic code of capitalism (p. 1038).

DECLINE AND FALL

In the latter days of the Empire, from the time of Hadrian (A.D. 117–138), the whole economy began to break down. The army, which had been a great source of wealth in slaves and loot, became an increasing but necessary burden, for now new lands were no longer being conquered and the Empire was finding its own defence increasingly difficult.* Attempts at reform only made things worse in the long run. Money economy was undermined by inflation and gave way to barter, based on exchange of goods largely locally produced and consumed. The *villas*, in which the rich took refuge to escape taxation, became centres of local production and gradually replaced the old cities as economic centres, and trade became more and more limited to luxuries. These were only the last symptoms of a disease that was inherent in the class society of the ancient world. There was no way of getting rid of exploitation short of a complete breakdown.

ECONOMIC AND INTELLECTUAL BREAKDOWN

Classical civilization was already intrinsically doomed by the third century B.C., if not earlier. The tragedy for science was that it took so long to die, because in that period most of what had been gained was lost. Knowledge that is not being used for the winning of further knowledge does not even remain – it decays and disappears. At first the volumes moulder on the shelves because very few need or want to read them; soon no one can understand them, they decay unread, and in the end, as was the legendary fate of the Great Library of Alexandria, the remainder are burnt to heat the public-bath water or disappear in a hundred obscure ways.

MYSTICISM AND ORGANIZED RELIGION

Thought did not stop with the fading away of natural science; it merely turned once more towards mysticism and religion. Though the emotional drive to mysticism is the desire to escape from this wicked world, it had an elaborate philosophic intellectual foundation, deriving from Plato at the time of the decay of the democratic city state. The subsequent schools, particularly the Stoics and the Neoplatonists, developed the mystical side of Plato's idealism and left out the mathematical, except

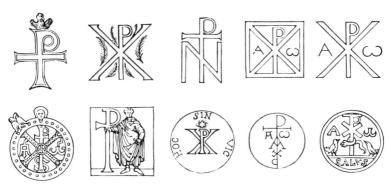

66. The earliest Christians were forced to keep their meetings secret and various secret signs and symbols gradually came into use. One of the more widespread of these was the chi-rho sign, made from a combination of the Greek letters chi (χ) and rho (ρ) which were the first two letters of the Greek word Christos ($\chi\rho\eta\sigma\tau\sigma$). These monograms are redrawn from those found in the very early Christian Church; the two on the right incorporate also an alpha and an omega – an allusion to the text 'I am the beginning and the end'.

67. Some late classical scholars took refuge in the Church, and the cross, the main symbol of Christianity, appeared in many forms. This Celtic cross at Iona dates from the tenth century.

in the form of a cabbalistic numerology abounding in magic squares and mystic numbers. From the first century onwards, philosophic mysticism fused with that of the salvation religions, of which Christianity was the most successful. Their common intellectual feature was a reliance on *inspiration* and *revelation* as a *higher* source of *truth* than the *senses* or even than *reason*: as Tertullian expressed it, 'I believe *because* it is absurd.'

The rise of these religions was itself a symptom of the hopelessness of the slave, and even of the citizen, in the face of a system that ground him down and from which it seemed impossible to escape (pp. 255 f.). He could take his choice of indulging in almost revolutionary denunciations of the system, such as are found in the Apocalypse, and stirring up resistance to official worship; or of retiring to the desert to avoid contamination with the evil of the world. To the religious it was not only idolatry but all that went with the hated, upper-class State that was abominable; the luxury, the art, the philosophy, the science were all signposts on the way to hell. Augustine and Ambrose, turning from wicked learning to holy nonsense, were just as much part of the movement as the monk-led mob who stoned Hypatia, one of the last of the Greek mathematicians. Only when the old classical world was utterly destroyed, as in the West, or tamed, as in the East, could the Church allow, and then very gradually and reluctantly, a limited secular science. How this happened will be told in the next chapter, which will trace the rise of the new civilizations which stemmed from the decay of the classical world. Here also will be found an account of Christianity, which though it arose out of classical civilization was a product of popular opposition to all that it stood for and properly belongs to the next stage of society. Despite its opposition to classical culture it would be absurd to blame Christianity for its decline and fall. It was a symptom rather than a cause. The mysticism, the absurdity, the confusion and decay of late classical times were the products of the social and economic collapse of the plutocratic slave State. In Aristotle's sense it was far gone in corruption; in the Chinese phrase, it had exhausted the mandate of Heaven. Although the rule of nominally Roman emperors in Constantinople was to last another thousand years, that empire belonged to a new age.

THE BARBARIANS

The final phases of the break-up of classical civilization took a different form in the older civilized and Hellenized eastern parts of the Empire than in the relatively recently conquered West, where city life was a foreign importation and the countryside was still largely pagan. The East absorbed its barbarians. City life never ceased and passed with hardly a break into the rule of Islamic Caliphs, and that of the (far more Greek than Roman) Byzantine emperors. The new structure of the States was not the same as the old, but trade, culture, and learning were preserved and for a while brilliantly revived.

In the West there was something like a general economic collapse of which the barbarian invaders took advantage. The barbarians were not themselves responsible for that economic breakdown. Far from invading the Empire, in the first place they were introduced into it as mercenaries, slaves, or serfs, largely to make up for the shortage of labour which the killing exploitation of the Roman landlords and tax-gatherers had already produced. Further, the Roman technique had not developed far in the practical field of food production in the heavily forested lands of the North and West. There seems to be no doubt that the barbarians themselves had better agricultural techniques than the Romans they displaced. At least they were able to cultivate the fertile and heavy soils of western Europe which the Romans neglected. In Britain, for example, the Roman estates covered only a fraction of the land occupied and effectively tilled by the heathen Saxons (p. 285).*

LOSS OF ORGANIZATION AND TECHNIQUE

What was lost in the barbarian invasion of western Europe was everything of culture that depended on large-scale material organization. Bridges, roads, aqueducts, irrigation canals, all fell into decay and largely disappeared. So did the distribution of standardized goods, such as pottery, from a few central factories. The only fine techniques to survive and flourish were those producing portable objects of fine metal-work for ornaments and weapons. With the disappearance of a literate class of wealthy people and their dependants in the cities there was little left of the tradition of philosophy, and hardly anything of science. Late classical scholars took refuge in the Church, like Gregory of Tours or Paulinus of Nola, or like Boëthius became officials of barbarian kings, or retired to their estates like Ausonius (*c*. A.D. 310–*c*. 395). Nevertheless enough was left in Europe of the classical culture to enable it to be reborn, purged of most of its limitations of the days of the Empire. In Venice, Salerno, and far-away Ireland were sources from which the fresh and original medieval culture was to flow, and to meet again in the twelfth century the main stream that had flowed through the Islamic East. [3.4; 3.47-9]

4.9 The Legacy of the Classical World

This book is concerned with the influence of science on history, and in particular with that of the natural science of the classical world on the life of the times and of succeeding ages. This chapter should serve to bring out something of what science meant and effected in the life of the Greek city. We are apt to be so dazzled by the intellectual and artistic brilliance of the Greeks that it is difficult to realize that their knowledge and skill affected far more the appearances than the practical and material realities of life. The beauties of Greek cities, temples, statues, and vases, the refinement of their logic, mathematics, and philosophy, blind us to the fact that the way of life for most people in civilized countries was, at the fall of the Roman Empire, much what it had been 2,000 years before when the old bronze age civilization collapsed. Agriculture, food, clothes, houses, were not notably improved. Except for a slight improvement in irrigation and road-making, and for new styles in monumental architecture and town planning, the science of the Greeks found little application. This is not surprising; for in the first place science was not developed by well-off citizens for that purpose, which they despised, and in the second, even with the best will in the world, the science they had acquired was far too limited and qualitative to be of much practical use. Greek mathematics, elegant and complete as it was, could be applied to few practical purposes for the lack of either experimental physics or accurate mechanics. The chief fruit of the magnificent Greek astronomy was, apart from astrological predictions, a good calendar and some indifferent maps. The great nursery of applied astronomy, the art of the navigator, hardly existed for lack of ships or incentives to sail the trackless ocean.

The other natural sciences were hardly more than discursive catalogues – such as Pliny's great *Natural History*[2.69] – of the common observations of smiths, cooks, farmers, fishermen, and doctors. Where science intervened it was to impose naïve or mystical theories, based on elements or humours, which confused and distorted the understanding of Nature. The consequences of the social sciences of the Greeks were more direct, though just because they were relative to the conditions of a city state they became inapplicable when these changed (pp. 1036 f.). The techniques, in contrast to the sciences, lasted far better and lost less. Indeed, except where they depended on scale, like the making of roads and aqueducts, they were transmitted unchanged in essentials, though, at least in the West, they were debased and simplified in expression.

The full possibilities of classical culture could not be realized in the framework of the civilization which gave it birth. They were blocked at every turn by the social and economic limitations inherent, as we have seen, in a slave-owning plutocracy. The real contribution of Greek science was to be in the future, though it could be made only in so far as the germinal elements of the classical culture could be preserved and

68. Travellers brought back tales from abroad, often embellished with imaginary incidents or perhaps with innocent misinterpretations of what they saw. Pliny the Elder (?33–79) was uncritical of such accounts which he mentioned in his *Natural History*. This sixteenth-century woodcut is of a phoenix, believed to be the only creature of its kind, with a life span of some six centuries, which burnt itself on a funeral pyre and emerged from its own ashes with renewed youth. From Conrad Lycosthenes, *Prodigiorum ac Ostentorum Chronicon*, Basle, 1557.

transmitted. Fortunately, though classical civilization had not the power to save itself, it had enough prestige to ensure that at least some of its achievements could never be forgotten and could later become the basis for new growth.

What had happened in the period of Hellenic and Roman power was a great spread of civilization all the way from the Atlantic to the Hindu Kush. The prestige which the extent of the power and the culture of these great empires generated far outlasted their political sway. It served, even after its original impulse was spent, to spread over a far wider area the ideas, the methods, the styles, and the techniques of Hellenism. In the East, Central Asia, China, and India all felt its influence blending with those of old native cultures; in the West the prestige of the lost learning served to tame the barbarians of Europe.

Indeed, perhaps the most important salvage of the Classical Age was the very idea of Natural Science. The belief persisted, as legends attest, that the Ancients through deep study had acquired a knowledge of Nature that enabled them to control it. Alexander, instructed by Aristotle, had a submarine and could fly through the air in an eagle-powered chariot. Of the actual elements of classical culture, science, particularly astronomy and mathematics, proved in fact the most lasting. Because they were needed to chart the planets, if only for astrological predictions, they had to be handed on and practised. Much of the other sciences was preserved in books, to be rediscovered at intervals by the Arabs and the Renaissance humanists. We shall never know how much was irretrievably lost, but certainly enough came through to guide and stir the thought and practice of later ages. So much, indeed, was rediscovered and imitated in the last 500 years that we have effectively incorporated the classical world in our own civilization, and nowhere more consciously or fruitfully than in technology and science.

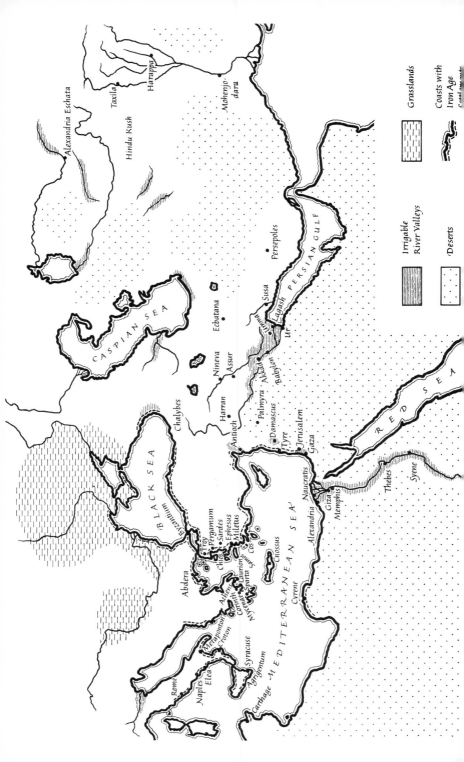

Grasslands

Coasts with Iron Age settlements

Irrigable River Valleys

Deserts

Alexandria Eschata

Taxila

Harappa

Hindu Kush

Mohenjo-daro

CASPIAN SEA

Ecbatana

Persepolis

PERSIAN GULF

Susa

Lagash

Ur

Nineveh

Assur

Akkad

Babylon

Harran

Palmyra

Damascus

Antioch

Tyre

Jerusalem

Gaza

Chalybes

BLACK SEA

Byzantium

Troy

Pergamum

Sardes

Chios

Ephesus

Miletus

Cos

Cnossus

Abdera

Argos

Athens

Corinth

Mycenae

Sparta

Laurion

RED SEA

Thebes

Syene

Giza

Memphis

Alexandria

Naucratis

Cyrene

MEDITERRANEAN SEA

Rome

Naples

Elea

Metapontum

Croton

Syracuse

Agrigentum

Carthage

Map 1

The Beginnings of Civilization

This map shows the major areas, with the exception of the Chinese plains, in which
we have evidence of the origin of agriculture and the building of cities. Most of the
area, apart from high mountains and deserts, consisted originally of open grass-
covered plains where pastoral culture could take form; the flood plains and deltas
of important rivers, which are suggested as the first localities for cities; and the
coastal areas opened up in the Iron Age. The localities of the principal cities of the
Bronze and Iron Ages are also indicated.

Table 1

The Development of Techniques and the Origins of Science (Chapters 2, 3, and 4)

This table shows the main technical developments from the period of the first human societies to the beginning of the classical period about 600 B.C. The dates are given only to indicate the beginning of the characteristic cultures of palaeolithic, neolithic, bronze, and iron, at their main centres of origin. Elsewhere they appear much later. In each period the arrangement is not chronological but merely a list of the most significant features of the stage of culture.

	Basic Food Production and Transport	Tools and Materials
The Palaeolithic Age Chapter 2	Food gathering and hunting	**Stone implements** Hand tools and weapons
	Organized big-game hunting Canoes Fishing, trapping Grain and root collecting	Hafted tools: hammer, axe, and spear Bow and sling Bow drill
The Neolithic Age Chapter 3.1	Agriculture Shifting hoe culture Domestic animals for food, wool, pack and draft use	Ground stone tools: axes, hoes Hand mills Rough carpentry
	Food storage Plough Permanent fields	Ornaments of native gold and copper
The Bronze Age Chapter 3.2–3.8	Irrigation Water-lifting devices Canals and dams Sail boats	**Metal** Mining and smelting Copper and bronze casting
	Wheeled carts Roads Horse chariots	Bronze tools, saws, chisels Weapons and armour Riveting, soldering, metal vessels
The Early Iron Age Chapter 4.1–4.3	Increased forest clearance and ploughing Waterwheels and pumps Gearing and pulleys Improved sea-going ships	**Iron** Improved and cheaper tools and weapons Catapults and other war machines

↓   ↘   ↓
Biology   Physics and Mechanics

Equipment and Processes	Social Organization	Intellectual and Cultural Achievements
Fire	Small social groups	**Language**
Cookery		Animal and plant lore
Roasting		
Prepared skins	Totemic clans	Ritual dancing, songs, and music
Clothes, bags, and buckets	Hunting rites	
Thongs and twine	Burial rites	Myths
Nets and ropes	Magicians	Naturalistic painting and
Baskets		sculpture
		Medicine and surgery
Spinning	**Villages**	Calendar for agricultural use
Weaving		
Reed and clay huts, wooden	Fertility rites	
houses	Rain makers and corn kings	Geometric design
Pottery		
Baking and brewing	Emergence of social	Symbolism
	differences	
	Ritual exchanges	Creation myths
Brick and stone building	**Cities**	Ideographic signs
	Class societies	Accounts
	Gods and temples	Numbers
Many-storeyed houses	Priest kings	**Writing**
Joined furniture	Craftsmen, traders, law,	Weights and measures
Chairs, beds, tables	property, and debt	Arithmetic and geometry
Beer and wine	City states and war	Solar calendar
	Empires and slavery	**Astronomy**
Glazed pottery	Barbarian irruptions	Professional medicine
Glass	Trading cities	Alphabet
	Politics	Literature
Improved preparation of	Republican government	
drugs and dyes	Rise of plutocracy	Coined money
	Social struggles	Philosophy
	Intensified warfare	**Birth of rational science**
↓	↓	↓
Chemistry	Social Sciences	Astronomy, Mathematics and Medicine

Table 2

Techniques and Science in Classical Times (Chapter 4)

This table covers the 1,100 years of the development of rational science, predominantly Hellenic, to bring out its relation to contemporary history and technique. The period is divided into centuries and as far as room allows individual contributions are attributed to the century in which they occur. No significance can be given to finer time intervals. The time-scale is uniform and the crowding of names in the Athenian and Hellenistic period brings out the great scientific activity in those periods compared with the comparative sterility of the Roman period.

		Technical Developments	Political and Social Events
	600 B.C.	Acquisition of Eastern techniques	Age of tyrants
Chapter 4.5	500	Mining and metal-working Shipbuilding Architecture and sculpture	Persian conquest of Ionia Greece liberated from Persians Pericles in Athens Peloponnesian War Athenian democracy
Ch. 4.6	400	City building on grid plan	Defeat and reaction in Athens Triumph of Macedon Alexander's conquests
Chapter 4.7	300	Geographical information on Persia and India Great development of water-works and military engineering	Hellenistic influence in Egypt, Persia, India, and Central Asia Carthaginian Wars
	200	Mechanical toys Great spread of slavery	Roman control of Greek world
	100		Roman civil wars Conquest of Gaul *Caesar* reforms calendar *Augustus* first Roman Emperor
	0	Spread of Roman architecture based on circular arch and vault	Jewish revolt Spread of Christianity
Chapter 4.8	100 A.D.	Water-mills	*Marcus Aurelius* the philosopher emperor
	200	Decline of city economy and trade	Crises and barbarian invasions *Diocletian* attempts to stabilize Empire *Constantine* Christianity official
	300		Condemnation of Arianism
	400		Breakdown of Western Empire Rome sacked by Goths *Augustine* 'City of God'
	500		Nestorian heresy

Philosophy and Science

Influence of Babylonian and Egyptian learning
Thales and the Nature Philosophers
Materialist theory of the universe
Heraclitus philosophy of change   *Pythagoras* number and form, physical law
Anaxagoras heavens not divine

                                           Philolaus spherical earth
Empedocles four elements   *Parmenides* change illusory
Hippocrates rational medicine   *Democritus* atomic theory

Socrates the dialectic method
Plato Idealism
Eudoxus heavenly spheres   *Aristotle* Reason and Logic, descriptive biology
                                           Theophrastus mineralogy
Museum of Alexandria   *Epicurus* atomic philosophy
Euclid ordered geometry
Strato experimental physics
Erasistratus human anatomy   *Aristarchus* rotating earth
Apollonius conic sections   *Archimedes* mechanics, hydrostatics
Ctesibius mechanics and pneumatics   *Eratosthenes* map and size of earth
                                           Hipparchus observational astronomy,
                                              precession of equinoxes

Cicero Greek philosophy for the Romans   *Lucretius* atomic materialism,
                                              science without religion

                                           Strabo geography
Pliny encyclopaedia   *Hero* mechanics, steam engine
Dioscorides descriptive botany   *Vitruvius* architecture

Galen codified medicine and physiology   *Ptolemy* 'Almages', descriptive astronomy

                                           Pappos calculation of areas and volumes
                                           Diophantus numerical equations

Zosimus, rise of alchemy, distillation

                                           Hypatia murdered

                                           Proclus last Greek mathematician

PART 3

Science in the Age of Faith

Introduction to Part 3

The period covered by this section is a vast one, ranging from the fading out of classical Graeco-Roman culture in the fifth century to the dawn of a new culture based on a new economic system and a new experimental science in the Renaissance. Nevertheless, for the objective of this book, the historical process over these ten centuries has a dynamic unity. Throughout the whole period we are seeing the decay, transmission, recovery, and the beginnings of the inner transformation of the body of techniques and beliefs that stem largely from the Hellenistic world. This holds not only for Europe but also for Asia, where (except for China, where a still older tradition remained dominant) technique and science had drawn deeply from the same source. The emergence of modern science is understandable only in terms of the Hellenistic world-picture epitomized in Plato and Aristotle. Throughout most of the period, indeed until well into the fifteenth century, the main intellectual task was one of recovering that picture and adapting it to the new, essentially feudal, economy that nearly everywhere accompanied the breakdown of that of slave-owning plutocracy. It was also necessary to adapt it to the cramping intellectual requirements of the dogmatic religion of Christianity, which had survived the breakdown of the old world and that of Islam, itself largely a product of that breakdown.

That this was possible at all, and that no radically new world-picture was needed, is an indication that the economy of feudalism, technically and economically more fragmented and primitive than that which it replaced, did not have great need for radically new intellectual forms and accordingly could not develop them. What it could and did do was to introduce new productive techniques which, though on a smaller scale, were far more widespread and closer to the people than were those of classical times. It was, as we shall show in Part 5 (Volume 2), this feature of later medieval life and the economic changes which accompanied it that gave rise to the radical transformation of the sixteenth century which at the same time created *modern science* and *capitalism*.

To explain the birth of modern science we need to know its antecedents, something of the long and very obscure period of preparation which led up to it, something of what it owes to the cultures of the classical and pre-classical civilizations as well as to those of Islam, Persia, India, and China. Most of all we need to know how it came about at all. What led to the appearance of the new science in the sixteenth century in Italy? What made it flower so abundantly in the England, France, and Holland of the seventeenth? Why had the same decisive steps not been taken in other cultures, such as those of India and China, which seemed ready for them at different periods of their history? These questions and some attempt to answer them form the central theme of this section. In it will be found an assessment of the factors that contributed to the rise of modern science. The most important of these are shown to be the economic tendencies, which in increasing measure throughout the later Middle Ages put a premium on technical advance, particularly in the direction of labour-saving. These are the same tendencies that mark the transformation of the economic structure of *feudalism* into that of *capitalism*. Indeed the track in time and place of the growth of capitalism in Europe is the same as that taken by the development of science. It will be shown how, in the early stages, science followed the development of nascent capitalism and how gradually it came to influence that development itself. The general character of science over the whole period was dictated by the existing feudal conditions which limited it rather than by anticipation of a different social state which was still to come.

The periods covered in Part 3 include that of the origins, the growth, the flowering and the decay of the feudal economy in northern Europe and the Mediterranean lands, together with the parallel but distinct developments in Asia, whose contribution to world culture was greatest of all in this time. They fall naturally into two very unequal parts. First, in Chapter 5, comes the transitional period of some 700 years, A.D. 450–1150, characterized in Europe by the salvage of a residue of classical techniques and science, and by their continued development in Syria, Egypt, Persia, India, and China, all under the impulse, direct or indirect, of Hellenistic culture. The results of all these were fused together towards the end of the period in *Islamic* culture, which, in its short but brilliant flowering, acted at once as the transmitter of the old culture and stimulator of a new advance in science.

The second period – that covered in Chapter 6, A.D. 1150–1440 – is a distinct one only in Europe. It begins in the field of science with the impact on a vigorous feudal society of the Islamic version of Hellenistic science, leading to the brilliant but unsustained movements of medieval scholasticism. It is also marked by a slow but accelerating movement of advance of techniques and scientific interests under conditions of an increasingly unstable *feudalism*. This advance in itself and in its economic consequences prepared the way for the next social form of *capitalism* in which *modern* science came into being, as will be told in Part 4 (Volume 2).

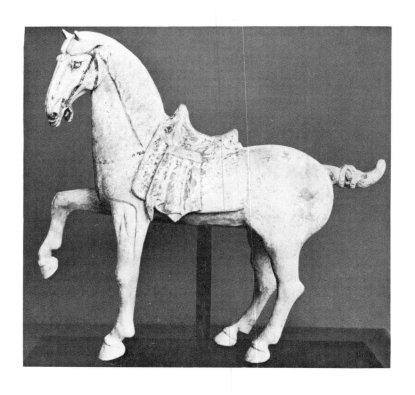

69. A beautiful example of Chinese ceramic art is this earthenware figure of a horse from the T'ang Dynasty (618–906).

Chapter 5

Science in the Transition to Feudalism

5.1 The Developments of Civilization
After the Fall of the Roman Empire

In our traditional education attention has been so fixed on the history of the Roman Empire, and particularly its western section, that we are apt to think that a general destruction of civilization occurred from the third to the ninth centuries. In fact all that happened was that in the most lately and artificially civilized parts of the ancient world, Britain, France, the Rhineland, Spain, and Italy, a system of government by a class of wealthy slave-owning patricians and provincials collapsed and was gradually replaced by a much more widely based, though incoherent, feudal order. The barbarian invasions that accompanied this change were its result and not its cause.

Meanwhile, over the rest of the Roman Empire, great cities such as Alexandria, Antioch, and Constantinople survived undamaged and orderly government, though increasingly restrictive, was maintained. Well beyond the bounds of the Roman Empire, over the whole territory which since Alexander's raid had fallen under Hellenistic influence, including Persia, India, and Central Asia, civilization continued to flourish and develop, but without the rigid economic, technical, artistic, and scientific limitations of late classical culture. The great periods of the Sassanian Empire in Persia (A.D. 226–637), of the Guptas (A.D. 320–480) and Chalukyas (A.D. 550–750) in India, and of the less known kingdoms of the Chorasmians in Central Asia (A.D. 400–600), all overlap the interval between the fifth and the ninth centuries that we call the Dark Ages, as if, because little is known of what happened in a very partially civilized western Europe, a great darkness covered the whole earth. Further still, China under the Wei (A.D. 386–549) and T'ang (A.D. 618–906) dynasties was enjoying a period of unexcelled economic and cultural achievements.[3.8]

In their economic and political structure all these States had not departed as far from the pattern of the early bronze age civilizations, which had existed in their territories, as had the cultures of Hellenized and Romanized countries. They had never undergone the intense economic and political struggles arising out of a money economy and slavery, which had first made and then destroyed classical civilization. In other respects their cultures were very different from each other. Persia was still dominated by an old tribal nobility and the simple religion of Zoroaster was being restored to vigour by a reforming dynasty. India had already, by the sixth century, developed the complex religious and caste system which Buddhism had been powerless to check, while China was well set along the path charted by Confucius with the dominance of a highly educated country gentry, though its culture still preserved many features of primitive clan society[2.34] expressed in the cult of ancestor worship.*

Though each culture followed its own pattern, they were at this period far more in contact with each other than before, particularly through the medium of trade. As a result of a wide market, though one limited to luxuries, manufacturing techniques improved, especially in weaving, pottery, and metalwork. The draw loom, irrigation machinery, and probably many of the key inventions in mechanics and navigation that were to transform Europe in the Middle Ages, arose in the East at that time. Art certainly flourished mightily, as the treasured objects of this period in our museums show. Although Hellenistic art was enthusiastically taken over as far as India and beyond, its cold ideal forms were rapidly transformed and given a new and sensuous life.

Of science we know little outside India and China, but we can infer from their rapid later flowering under the protection, but not necessarily under the impetus, of Islam that it was as much cultivated in Persia and Central Asia. Greek influence is visible particularly in mathematics, astronomy, and medicine, but transplanted to a new medium it could grow in a way that it could no longer do in its own country. All these developments were later to contribute to a common cultural advance, but they are not by themselves as important as the basic economic changes which accompanied them.

The decline and fall of the Roman Empire marked a definitive era in the history of the whole of humanity. In its prime it had been the largest State in the world. Its military and civil organization and its trade had reached the limits of size of any human community for many centuries to come. None of the States which took its place over its old territories ever managed to maintain such an organization for such a time over so large an area. Outside it the only comparable empire was that of China, and the character of Chinese State organization was very different from the classical one. As the Roman, plutocratic, slave economy disintegrated, for reasons already discussed, it left behind it almost everywhere the seeds of a new decentralized economic and political system.

There are marked resemblances but even greater differences between the immediate consequences of the Roman collapse and that of the old bronze age civilization 2,000 years before (p. 143). In both cases life took up again from a lower technical level, but in the later case the relative economic fall was even greater, at least in Europe. On the other hand, as we shall see, much more was salvaged of knowledge and culture. What disappeared, as in the earlier case, was everything dependent on large-scale operation, communications, long-range trade, waterworks. What took the place of the Roman Empire was, however, something quite different from the swarm of trading and ultimately democratic city states that marked the beginning of the Iron Age.

THE TRANSITION TO FEUDALISM

Despite the continued existence of cities in the Eastern Empire the economy of the new order was everywhere essentially country based, the unit being the estate, villa, or manor worked by serfs, rather than chattel slaves, who were permanently attached to the land with rights to compensate for their heavy duties. The estates were owned either by the descendants of the old city plutocracy, as happened mostly in the Eastern Empire, or by barbarian clan chiefs in territories occupied by Germans or Arabs. The economy of the countryside was essentially feudal both in the lands of the East, where the owners at first lived largely in the towns, and in the West with its poorer communications where they lived on their estates.

In most cases the peasants, *coloni*, serfs, *rayats*, remained in possession of the land and tools, but were forced to yield part of the produce or labour to their lords in the form of rent, taxes, or feudal service. The standards of land use reverted in the West to a subsistence economy, but one on a higher technical level than that of the Iron Age. In the East a larger surplus always remained for trade. The transformation to feudalism naturally did not occur all at once, it took several hundred years, nor did it proceed at the same speed in different places. Before feudalism reached its full extent it had already begun to decay at the centre. Nor was it limited to the areas of the old Greek or Roman Empires. As the predominant economic mode it spread as new land was opened to cultivation in Europe and Asia.

5.2 The Age of Faiths

The conditions of feudal production reduced the demand for useful science to a minimum. It was not to increase again till trade and navigation created new needs in the later Middle Ages. Intellectual effort was to go in other directions and largely in the service of a radically new feature of civilization – *organized religious faiths*.

The advent of organized religious faiths as a dominant political and social force, which occurred in the earlier centuries of our era, was a development by no means limited to Christianity. It was a world-wide phenomenon, showing many similar features in widely different regions, and indicating that it arose from a common need by virtue of common possibilities. Between the third and the seventh century A.D. we find the rise to power and influence of Christianity, of Islam, and of Buddhism in China and south-east Asia. Buddhism in India and Zoroastrianism in Persia were, it is true, founded as religions some seven centuries earlier, but it was in this period that their doctrines were fixed and their priest-hoods organized. This was also when even the most multiform and unorganized religion, Hinduism, which was replacing Buddhism in India, established itself anew and codified its sacred books.*

It would seem as if, for the first time in human history, there was the need for religions based on a fixed system of beliefs together with the means of maintaining them. A clue to the explanation of the latter condition is given by some of the characters of organized religion, which in various degrees are found in all, or nearly all, of them. They are a hierarchic priesthood, fixed rituals, and as a test and rallying point a *creed* involving belief in an order of the universe, embodied in *sacred books*. In addition there are auxiliary features which are more variable – the appearance of devotees – either singly as hermits, fakirs, yogis, or in bodies as monks, lamas, or dervishes – given over to asceticism, begging, preaching, or occasionally working. Certain of these practices are far earlier than organized religion, and indeed are found in the most primitive communities, but they took on a new aspect in relation to advanced city life. Hermits and monks represent the religious side of the flight from the oppressive and sinful cities in the period of their decay, the secular side being the retirement of the wealthy to their country estates to evade the imperial tax collectors.[3.4]

The central feature of the new, organized religions is the social coherence of the Church and the creed it defines and imposes. It lies in common ritual and common philosophic beliefs. The fact that these religions are all, in Mohammed's phrase, 'peoples of the book', shows that they imply a certain degree of literary culture in a numerous if restricted class. The fact that the ritual and ministrations of the Church are extended to all the people shows that at the same time the priesthood aim at securing a universal or *catholic* assent. The new religions were indeed, once they had outgrown their formative revolutionary phases, essential stabilizing organizations. They aimed, often unconsciously but sometimes consciously, at making the social order generally acceptable by showing it to be an integral part of an unchangeable universe (p. 1042). At the same time the introduction of gods, myths, and visions of a future life provide distractions and a celestial balance to the injustice of this world.

EARLY CHRISTIANITY

These features are particularly evident in the history of early Christianity. A knowledge of this history is of unique importance to the understanding of science, for it was within the framework of Christianity, except for a brief period under Islam, that modern science grew to maturity. Christianity arose out of the distress and aspirations of the common peoples of the Roman Empire (p. 233). It is no accident that it first appeared among the Jews, who were, if not the most oppressed, certainly the most rebellious of the subject peoples. Jesus himself, as a hoped-for Messiah, was taken for a revolutionary and suffered a revolutionary's fate. The early Christian communities were themselves, or were closely modelled on, those of the Essenes.[3.56] These had been formed as closed, economically self-supporting, communist groups of Jews who rejected both the compromises with wealth and foreign customs, into which the originally revolutionary Maccabees had been betrayed (p. 158), and the ritual particularism of the Pharisees.*

This association with the Jewish democratic tradition and especially with the rejection of any compromise with the powers of this world assured early Christianity of popular support which was only reinforced by official persecution. The popular appeal of Christianity was at its greatest in its first two centuries, just at the time when the Empire seemed most safe and glorious to the wealthy and cultured citizens. It was then that Roman rule bore most hardly on the common man and the slave. For them there was no hope in this world and little reason to dread its fiery end. Christianity was able to spread far more widely than Judaism because it shook itself free from the tribal particularism of Judaism while preserving all of its popular appeal. It was something far more than just another mystery religion such as Mithraism, which also flourished mightily in that disturbed time. Christianity furnished an all-inclusive organization that, however outwardly submissive, was absolutely determined to have no part in the oppressive and sinful classical civilization. Inevitably it became a political movement, representing at the outset the aspirations of the oppressed lower classes in the great cities and a national reaction of the oriental peoples against dominant, upper-class Hellenism.

Christianity did not, however, long remain confined to the lower classes, and little by little, as it came to include more and more cultured proselytes, many of the ideas of the classical world crept into its teaching. Some were much more easily assimilable than others; in particular, Platonism and, even more, its half-Christianized offshoot, Neoplatonism, which was so useful in emphasizing the 'other-worldliness' of religion. The two aspects – the popular revolutionary apocalyptic aspect of religion, with its vision of a Last Judgement and a Kingdom of God in our time; and the other-worldly spiritual attitude, very much more favoured by the upper classes – have run through the whole of Christian history to this day.[2.74]

It would be wrong now, however excusable in Gibbon's day, to hold Christianity, as such, to blame for the economic or cultural collapse of classical civilization. The causes of this, as has been shown earlier, were intrinsic. The Church, which was to play the dominant role in the subsequent Dark and Middle Ages, did determine, to a large extent, the character of the culture that it installed in its place. The Church was the one coherent institution of the late classical world that had survived the troubles of the fall of the Empire in the West. It had also, long before that fall was complete, penetrated far beyond the ancient bounds of the Empire to cover much of Europe from Ireland to the Caucasus and had spread widely into Asia. To an extent unparalleled since the days of

70. The concept of the Last Judgement has run through Christianity since its inception. This carving appears on the tympanum of Bourges Cathedral.

ancient Egypt, culture and even literacy were confined to the clergy. The Church, in addition to its spiritual functions, provided for education, administration, and, in the early Middle Ages, for law and medicine as well.

ECCLESIASTICAL ORGANIZATION

It was no accident that the Church survived the Empire; it had far more solid political and economic foundations. Beginning as a virtually revolutionary movement – true, with an other-worldly objective, but nevertheless openly antagonistic to civil administration – it early acquired in self-protection a close organization, part agitational, part economic. This organization at first through its elders – *presbuteroi*, priests – and their servants – *diaconoi*, deacons, deans – kept in personal touch with every individual Christian and could count on his support in a way no imperial official could hope to do. Later in the second century, as the Church grew in numbers, higher organization was necessary to ensure that doctrinal and personal quarrels did not split it into innumerable fragments. A parallel organization to the State was built up, often using the same terms such as *ecclesia* – église – church, *basilica* – royal palace, and *diocese*. Inspectors – *episcopoi*, bishops – were ordained, and later the most important of these became the great *patriarchs* of Jerusalem, Rome, Constantinople, Alexandria, and Antioch. Centuries passed before the bishops of Rome claimed the primacy as the Holy Father, the Pope, God's vicar on earth, the Pontifex Maximus or Chief Bridge-builder – once merely across the Tiber, but now between heaven and earth.[3,4]

By the third century the Christian Church, though it still included only a small minority of the population, was the most powerful, widespread, and influential political organization in the Empire. Desperate persecutions failed to break it. By the fourth century it became clear that the only way to save the Empire was to take the Church over, and Constantine, long before he became a Christian, took this final step in 312 A.D.

THE END OF PAGANISM

Once the Church was in power, and disposing at the same time of patronage and punishments, the pagans, at least in the towns, were soon won over. There was, in any case, little resistance. The worship of the Olympians was by then not very serious and had only a snob value. As for philosophy, almost every school could be found in Christianity itself. What the Church could still not tolerate was any philosophy officially independent of Christian revelation. But it did not, however, usually suppress it directly. The murder of the mathematician Hypatia was not policy, but monastic zeal getting out of hand. More typical of the end of classical science was the closing of the schools of Athens by the great Christian emperor, Justinian, in A.D. 529. The last of the professors were allowed to go to the new university of the Persian emperor Chosroes at Jundishapur (p. 262). They found this atmosphere too strange and the Emperor sent them back with a treaty stipulation that they should not be molested.

More significant for the future was the conversion to Christianity of the philosopher now known as John Philoponos (*fl.* A.D. 530), which occurred about the same time. The conversion was whole-hearted; on going over he joined a kind of Christian Action party in Alexandria, the 'Philopoenes' or 'trouble lovers', mainly occupied in 'fighting against pagan professors and from time to time attacking the last temples of the Egyptian gods'. In the end he went too far and became a hyper-trinitarian, a tritheist heretic. In his rejection of pagan philosophy, Philoponos even had the temerity to deny Aristotle's theory of motion and founded the doctrine of 'impetus' which, after attracting some support from the Arabs and schoolmen (p. 301), was in the hands of Galileo to lead to the emergence of modern dynamics (pp. 428 f.).[3.10]

5.3 Dogma and Science

The triumph of Christianity effectively meant that from the fourth century onwards in the West, and up to the rise of Islam in the East as well, all intellectual life, including science, was inevitably expressed in terms of Christian dogma and, increasingly as time went on, was confined to churchmen. Between the fourth and the seventh centuries the history of thought over the area of the vanishing Roman Empire is the history of Christian thought.

In the early days of Christianity, science and learning had been associated with the hated pagan upper classes and looked upon with suspicion. But this attitude did not last. The human message of Jesus could hardly suffice the Church once it aspired to cultural pre-eminence. As the Gospel of St John shows, with its cult of the divine word – the logos – mystical Platonism was already at work in the foundations and indeed, in a more diluted form, it is already evident in the message of St Paul.[2.74]

71. The influence of dogma and mysticism on science was a powerful factor for many centuries. Some teachings from the Jewish cabbala and belief in the superiority of ancient Egyptian knowledge, particularly that of the fictitious Hermes Trismegistus, coloured even seventeenth-century science, as this title page shows. From the Jesuit, Athanasius Kircher, *Ars Magna Lucis et Umbrae*, Rome, 1646. Kircher's book contains much sound experimental evidence about the behaviour of light, yet is shot through and through with fables and mystical interpretations which were later to be declared heretical.

ORTHODOXY AND HERESY

The fathers of the Church, particularly Origen (*c.* 185–253), a school-fellow of Plotinus the founder of Neoplatonism, set about incorporating the safer parts of ancient philosophy into Christian dogma. Much of it had already found its place there unconsciously. The task was nevertheless difficult, partly owing to the very different philosophy which underlies the Old Testament (pp. 157 f.). Inevitably it led to controversies in which each side claimed to be orthodox and accused the other of heresy. The great disputes and heresies of the fourth and fifth centuries which split Eastern Christianity, those of the Arians, the Nestorians, and the Monophysites, were largely on points of interpretation of Neoplatonist ideas of the nature of the soul and its relation to corruptible or incorruptible bodies.

These disputes were nominally settled by Councils of Bishops, implying a basic democracy in the Church, but usually the decision went in favour of the side that could win over the Emperor. The great Arian heresy of the fourth century on the nature of the Godhead was settled in this way at the Council of Nicaea in 325. There Athanasius imposed his implacable Trinitarian creed. Its triumph was not assured, however, until almost two centuries later when Justinian had defeated the Arian Goths.

By the fifth century a compromise between faith and philosophy was worked out by St Augustine (354–430), who produced a kind of composite between scriptural tradition and Platonism, with a strong flavour of predestination, derived from his Manichean experience (p. 268), which was to dog Christianity and particularly Puritanism ever afterwards. This included the essentially Zoroastrian idea of the cosmic conflict of good and evil (Ormuzd and Ariman) with its associated ideas of the Devil and Hell-fire. The Augustinian compromise did not last; heresy followed heresy, and the work of suppressing them had to be done all over again in the Middle Ages (p. 294), and ultimately failed altogether in the Reformation.

The philosophies on which theology was based, though subject to dispute, were all readily assimilable to an other-worldly religion, while the sciences of observation and experiment were not. In the first place these were plainly unnecessary to salvation, in the second, by the mere dependence on the senses, they depreciated the value of revelation. The overcoming of this attitude was to be the work of many centuries, and was only to be achieved in an economic and social atmosphere very different from that of the decaying Roman Empire.

In all these religious disputes natural science was a certain casualty.

72. The canon of the Scriptures accepted by Western Christendom was due in a large measure to the work of St Jerome (*c.* 340–420) who translated the Hebrew 'Old Testament' into Latin and so provided a more accurate translation than was previously available, and who also translated the New Testament. His version of the Scriptures became known as the 'Vulgate'. This painting of the Saint in his study is by Vincenzo Catena (?–1531).

Classical philosophy, especially in its latter days, was absurd enough. The Old and New Testaments were never intended as interpretations of Nature. They contain moreover mythical and philosophical interpretations of all ages from the most ancient Babylonian onwards, and are therefore intrinsically self-contradictory.[2.74] To attempt to combine philosophy and scripture is a task defying all reason, and fatal to any clear understanding of Nature. Faith and reason cannot be reconciled without allegorizing the one or distorting the other – in either case discouraging honest thinking.

It is fashionable in these days to praise the Church for preserving the science of antiquity down to our times. The survival of science, as will be shown, has been due rather to its success, where faith failed, in coping with the real world. It has survived in spite and not because of the centuries of effort to subordinate it to out-dated and contradictory beliefs. As we shall see in case after case right down to the controversy on Darwinian evolution (p. 662), the acceptance of obvious solutions has been held up for scores of years because they could not be made to square with Genesis. To say this is not in any way to blame the Church or the clerics, who in their time did their best according to their lights, but only those who today ought to know better. If science advanced slowly in Christendom until the time of the Renaissance it was primarily not because of the Church, but because of the economic conditions that maintained it so long in its obscurantist role. Under feudal conditions advance could not have been faster.

5.4 The Reaction to Hellenism

SCIENCE IN SYRIA AND EGYPT

The Arian heresy was followed by many others. Two of these, however, those of the Nestorians and Monophysites, are of particular importance because they gave a decisive impetus to a national anti-Hellenic movement in Egypt and Syria, because they helped science to spread throughout Asia, and because they paved the way for the triumph of Islam. Once Christianity had become the official religion of the Empire, latent national or regional independence movements were bound to rally round heresies. What the heresies were is not now a matter of great moment. In A.D. 428 the Syrian monk Nestor maintained that Mary should not be called the mother of God, as she was only the mother of the human and not of the divine nature of Jesus. He was condemned at the Council of Ephesus (A.D. 431) and thousands of Syrian clerics, monks, and laymen faced persecution in his support. In doing so they defied the hated Byzantine government, and asserted their dormant Syrian nationalism against the Greek officials and upper class. The persecution was too effective to be resisted within the bounds of the Empire – and many Nestorians crossed the boundary into Persia, where a vigorous culture was being promoted by the Sassanian kings. Despite the official Zoro-astrianism, they were well received on account of their medical and astronomical knowledge, and were established near the King's court at Jundishapur, where they built a famous observatory. Nestorian monks penetrated the whole of Persia, made converts, and set up churches as far away as China.

Sixteen years later Eutyches of Alexandria (378–454), in his desire to avoid Nestorian heresy, went so far as to declare Christ's human and divine nature to be one and the same. This one-nature – Mono-physite –

heresy was duly condemned by the Council of Chalcedon (A.D. 451) under imperial pressure. Virtually the whole of the Egyptian clergy and many in Syria and Asia Minor defied the ban. Christians in Egypt and Abyssinia remain Monophysite to this day.

Persecuted Monophysites fled to Persia and quarrelled with Nestorians there. They too shook the dust of Hellenism off their feet and built up a vernacular Syrian science for theological purposes. This involved translating major Greek philosophic works into Syriac and thus starting the first independent national offshoot of Greek science.[3.43] These developments coincided with a vigorous economic upsurge in Syria which carried Syrian merchants, as successful rivals to the Greeks, all over the Mediterranean, and as far as Britain as well as over large parts of Asia.

THE FLOWERING OF INDIAN CULTURE

For the 500 years that followed the collapse of Rome the centre of science was shifted to the east of the Euphrates. The fifth, sixth, and

73. The Hindu temple was of great architectural elegance and this photograph of the rock-cut temple at Ellora is an excellent example. It was built in the second half of the eighth century A.D. by Rāshtrakūta, king Krishna I (Akālavarsha), although repairs and stucco reliefs on the exterior are of the late eighteenth century.

seventh centuries were an age of great cultural advance not only in Persia and Syria but also in India. Under the protection of the vigorous dynasties of the Chalukyas and Rastrakutas, an effete Buddhism was replaced by a renaissance of Hinduism, to which the magnificent temples of Elephanta and Ellora bear witness. There was also, and this is of the greatest importance for the whole world, a new development of science, particularly mathematics and astronomy, associated with the names of the two Aryabhatas and Virahamihira in the fifth century, and with Brahmagupta in the seventh. The basis here was Hellenistic science with some additions directly from Babylonia [2.62] and probably also from China.*

HINDU NUMBERS: THE ZERO

A decisive new development was made there about this time: the perfection of a *number system* with *place notation* and a zero – our modern so-called Arabic *numerals*, which made computation something any child could learn. It is significant that its first mention in the West was in 662 by Severus Sebockt, a Monophysite bishop in Syria. Another Syrian, Job of Edessa (*c.* 800), in a very fanciful style, after equating the nine digits with the nine choirs of angels (p. 308), explained the reason for the roundness of the zero in these terms:[3.38]

> The movement of numbering is completed in a kind of cycle. It is for this reason that the Ancients invented, as a first sign for this number [ten], the [empty] space between the forefinger and the thumb, formed in a circular way.
>
> Indeed when the numbers which we have with us reach a denary state they stop, and then turn back and mount up indefinitely.

Elements of Hellenistic culture, including science as well as art, penetrated in this period with Buddhism to China and even to Japan. There they blended with a still evolving old Chinese culture, whose contribution to the main stream of technology and science was, however, to come somewhat later (pp. 311 ff.).

THE CULTURE OF BYZANTIUM

Taken all in all, the sixth and seventh centuries, far from being the darkest of the Dark Ages, were a period of a growing world-wide civilization in which the heritage of Greece was everywhere engendering new beauty and new thought. This holds with limitations even for the surviving, and by then almost completely Greek, Eastern Empire of Constantinople. There, under emperors like Justinian (*c.* 482–565), there was a great revival in arts and techniques, as witness the mosaics and architecture of

74. The interior of the mosque of St Sophia, showing its fine Byzantine architecture. Built under Justinian (see plate 308, Volume 4) as a cathedral and dedicated to St Sophia, it became a mosque in 1453.

St Sophia. But although the tradition of Greek philosophy and science was preserved in the Byzantine culture, it lacked the power of growth. This was in part due to clerical obscurantism – it was in response to it that Justinian had closed the schools of Athens – but far more to the fact that the Greek tradition on its home ground was dead. It was old stuff, respected but not exciting, and it bore no relation to the current realities of monastic rivalries, palace intrigues, and the chariot races in the hippodrome.

THE TRANSMISSION OF CLASSICAL CULTURE

The breakdown of classical civilization, like that of the old river civilizations 2,000 years before, was by no means an unmitigated disaster to science. The new civilization that gradually replaced it escaped some of the limitations which had previously choked the progress that had started so hopefully in early Greek times. The two transitions, however, differ in one very important factor. Whereas there was little conscious continuity and no feeling of parentage or respect between the culture of the early civilizations and that of Greece, there was between classical culture and that of Syria, of Islam, of medieval, and still more of Renaissance Europe a continuity based on written documents and a strong feeling of being the heirs of the Ancients. The main thread of tradition had indeed never been lost; throughout the Middle Ages, Muslim and Christian alike had access to the works of many of the major thinkers of classical times. These works, as well as many others, were made available to a far wider audience in the Renaissance through the medium of printing.

It would be a mistake, natural enough in the time of the Renaissance but unpardonable now, to assume that all that happened then was the taking up again of classical culture where it left off, or even where it was at its best. What happened was something different and far more important. The civilizations that took over the classical heritage of science had a hard task to prevent themselves from being stifled by it. We have seen in the last chapter the low state of activity into which it had fallen even in the East. There was still, however, the vast store of knowledge to be found in books available to any with the desire or skill to read them. The Syrians and Arabs, and after them the medieval schoolmen and the humanists of the Renaissance, had to trace that store step by step back to its Greek originals, resisting as well as they could the temptation to accept what they did not understand as the holy and mysterious knowledge of the Ancients. That they managed to absorb and transform it at all was by virtue of their own vigorous cultural developments. The very rediscovery of the works of the Ancients was the effect, far more than the cause, of the spurts of intellectual activity that characterized the beginning of Islamic science in the ninth century, of medieval science in the twelfth, and of Renaissance science in the fifteenth century.

These advances were the easier because at each stage the new knowledge covered a much wider field of interest than the old. Late classical culture was limited both socially and geographically. Socially it had become an almost exclusively upper-class preserve and was consequently abstract and literary, for ingrained intellectual snobbery had barred the learned from access to the enormous wealth of practical knowledge that was locked in the traditions of almost illiterate craftsmen. One of the greatest achievements of the new movement which culminated in the Renaissance was to raise the dignity of the crafts and to break down the barriers between them and the learned world.

The geographical range of classical culture had largely been limited to the countries of the Mediterranean and the Near East. Its very completeness formed a barrier to the use of the common stock of techniques and ideas of the other ancient cultures of India and China. With the breakdown of the Roman Empire the way was open to much wider exchanges and influence.

5.5 Mohammed and the Rise of Islam

To these negative factors of release there was soon to be added a positive one – the appearance and rapid spread of a new world religion. The barriers of language, religion, and government that up to the seventh century had limited each culture to its own region were suddenly swept away over nearly the whole area of the ancient civilizations, stretching all the way from the Indus to the Atlantic. The advent of Islam, though determined in its particular form by the personal character of Mohammed, was by no means an inexplicable or even entirely unique phenomenon. The decay of the power of the Roman Empire did not affect its prestige, which long survived it, still less the influence of the popular religion of Christianity which gradually came to dominate it, and which spread further than that of its Church and creed. Nevertheless, unlike northern Europe, where no other culture had been known and where Roman power had long since lost its terrors, the peoples on the eastern fringe of the Empire were reluctant to adopt Christianity as too identified with an alien, hostile, or oppressive government. At the same time neither the official Zoroastrianism of Persia nor the local gods of Arabian and African tribes could compete with the intellectually coherent and emotionally stirring content of Christianity. The way was open for the formation of new synthetic prophetic religions, popularly based, and incorporating as much of Christianity as could easily be acceptable without submitting to its Church or accepting its doctrine.

The first of these attempts, the mission of Mani in the third century, had a lasting but limited success. Mani claimed to be the third and final prophet following Zoroaster and Christ, and carried a message of eternal salvation for the *predestined elect* and consolation in this life for the faithful who ministered to them. Mani was martyred in *c.* 276 and his followers persecuted in Persia, but their influence spread as far as China in the East and Provence in the West, and some of their doctrines, especially that of predestination, entering Christianity through their most eminent convert St Augustine, were to reappear again in Calvinism (p. 384).

The mission of Mohammed between 622 and 632, arising among the already vigorous and expansive Arabs, who only had to face the weakened and divided Roman and Persian Empires, had a greater promise of success. It still remains an almost incredible achievement for one man. Mohammed swept away the old tribal gods and replaced them by one God, Allah. Islam made a brotherly appeal to all men, it had a simple but exacting personal ritual, a theology reduced to bare monotheism, and it gave a sure hope of a realistic paradise for the believer. All this was conveyed in a poetic book, the Koran, which was not only an inspiration but a manual of ritual, morals, and law. It commanded then and still commands the devotion of poor and rich alike.

There was in Islam no church or priests, only the need for a court – Musjib (Mosque) – for common prayers and for readers of the Koran – *imam* – who were at the same time preachers and expounders of the law. Islam was from the beginning a literate religion. The Koran is still the common text-book for all Muslim peoples. The Caliph was the revered successor of the Prophet, at first also a civil ruler, but the strength of the religion did not lie in authority, but in the widespread religious community of the faithful. The political evolution of the early religious kingdom followed at first the late Roman or Byzantine pattern of a wealthy and luxurious court, torn by intrigue and depending increasingly on a praetorian guard of foreign, usually Turkish, slaves. This led to a break-up of Islam after its first two centuries into more and more feudal principalities, which were to be an easy prey to the nomads of the great plains and even to ill-organized and quarrelsome Crusaders. The religion of Islam, on the other hand, was solidly based in the people and was to outlast all misrule and conquest. Even as Christianity did in the North, it was to convert its conquerors, and was to be spread over a large part of Asia and Africa, where it maintained a coherent culture which, though not progressive, has persisted to our day.

The rise of Islam was abrupt. Within five years of the death of Mohammed in 632 the armies of his followers had decisively defeated both the Roman and Persian armies. After that there was for many years to be no force that could resist them. By the eighth century they had extended their conquests from Central Asia to Spain. The Roman dominions in Africa and Asia, with the important exception of Asia Minor, were in their hands as well as the whole empire of Persia, stretching right over Central Asia and into India. From that time on most of this vast area was to have a common culture, a common religion, and a common literary language. For some centuries it was to have a common government and free trade. For even longer religion and the pilgrimage ensured free passage from Morocco to China of scholars and poets.

THE ARAB RENAISSANCE

The immediate effect was a great stimulus to culture and science. The Arabs were no strangers to civilization. They had their own cities and had fulfilled an essential function in organizing the eastern trade of the Roman Empire. The ease of their conquest showed that all they did was to take over the urban civilization of the Mediterranean with the effective consent of the inhabitants. By that time few of these were prepared to fight to keep up an imperial government which did little but impose heavier and heavier taxation for increasingly ineffective services. The fact that Christianity was now the official religion hindered rather than helped the resistance of the populations of the Asian and African parts of the Empire, who were largely heretics and were safer from persecution under the Muslim Caliphs than under the orthodox emperors.

The Arabs, apart from securing for themselves the revenues of magnates and officials, were not in the least inclined to interfere with the local or city economy. The whole of the administration of the Omayyad Caliphate of Damascus was carried out by Greek officials in Greek. There was accordingly no specifically Islamic economic system. It was simply a late classical urban economy with the military command reserved, at first, for pure-blooded Arabs, but later falling, as in Rome, into the hands of any effective adventurer. Slavery did not disappear but, for lack of supplies of slaves, it was largely reduced to domestic service. Where there was gang slavery there were mass revolts, and that of the negro Zanjs from the saltpetre works of the Persian Gulf proved as formidable as the Spartacists of Roman times. The land was tilled by heavily taxed *rayats*, virtually serfs. These also often rebelled. One such rebellion, that of the communist Karmatians, maintained itself for over 100 years.

With reviving trade, merchants became relatively more important than in late classical times. Indeed the unity of Islam greatly helped trade by restoring the wide sphere that the Roman Empire had lost in the troubles of its latter years, and at the same time by extending and decentralizing it. In the whole area of the Muslim conquest, from Cordoba to Bukhara, there was no one centre which, like Rome, dominated and sucked in the economy of the Empire. Mecca was always a religious, not a political, economic, or cultural centre. Instead, not only did old cities, such as Alexandria, Antioch, and Damascus, take on a new lease of life, but also new cities on the same model appeared everywhere, particularly the great new capitals of Cairo, Baghdad, and Cordoba. All these cities were in constant communication with each other, and their varied products formed a basis both for trade and technical improvement.

Further, the cities of Islam were not isolated from the rest of the eastern world, as had been those of the Roman Empire. Islam became the focal point of Asian and European knowledge. As a result there came into the common pool a new series of inventions quite unknown and inaccessible to Greek and Roman technology. These included such manufactures as steel, silk, paper, and porcelain. In turn these formed the basis for further advances, which were able to stimulate the West to its great technical and scientific revolution of the seventeenth and eighteenth centuries.

THE REVIVAL OF CLASSICAL SCIENCE

On the intellectual side also there was very little break in continuity. The religion of Islam had at the outset, though not later, far less cramping effects on human thought than that of Christianity. By the time it appeared there was no danger to faith in paganism or philosophy. After the turbulent century of conquest even the leaders of Islam sought avidly for the old knowledge of the Greeks, and as much of their other culture as the Koran would allow them.

This impact of foreign influences coincided with the fall of the Omayyad dynasty of Damascus and the advent to power in A.D. 749 of the Abbasids, who, though not themselves Persian, depended on Persian support and liberated the traditional learning and science of that ancient and cultured people.* Learned Persians, Jews, Greeks, Syrians, and a few from farther lands met in the new capital of Baghdad. It was there and in Jundishapur that began the translation into Arabic of the main books of Greek science. [3,43] These translations were made either directly from the Greek or, more often, from the Syriac, and the work was subsidized from the start by the Caliphs and notables. Caliph al-Mamun actually founded a bureau of translation, Dar el Hikhma, where the great scholars Hunain ibn Ishaac and Thabit ibn Khurra produced Arabic texts of most of Aristotle and Ptolemy. They also translated many Persian and Indian books, but these were not further translated into Latin and were thus lost to the West.

The books that were translated were nearly all of science and philosophy because, naturally enough, the Arabs had no particular interest in the history of the Greeks. As for Greek drama and poetry, this had relatively little to give to a people who had a rich source of legend and a living poetry themselves. It was largely as the result of this concentration of interest that when in turn Islamic knowledge came to be transmitted to the West, it was at first limited to science and philosophy. The humanities were for the most part rediscovered directly from the Greek and Latin authors only in the Renaissance. The fact that the sciences and the humanities entered modern culture by such different channels is an important factor in the development of science, and has had much to do with setting up the barrier between science and the humanities that persists to the present day.

5.6 Islamic Science

It is difficult to estimate the value of the actual contributions to this fund of learning that were provided by Islamic scholars themselves. Certainly the learning of the Greeks was brought to life again and not merely transmitted without change. In fact it was subjected to a process similar to that undergone by the learning of the ancient East in the hands of the Greeks, though in this case the affiliation was far more direct and acknowledged. Because the Islamic scholars had no emotional identification with the old legends of the Greeks, they approached Greek learning with a much more detached attitude than the Greeks themselves were able to do. On reading Islamic scientific works one is struck by a rationality of treatment that we associate with modern science. On the other hand the Muslims were equally, if not more, attracted to the mystical aspects of late classical philosophy, particularly Neoplatonism, which they at first were unable to distinguish from Aristotle, owing to the incorporation in his works of such forgeries as *The Theology of Aristotle* and *The Secret of Secrets*. Much of this mystical confusion was passed on through them to the medieval schoolmen. Another misfortune that was to dog not only Islamic but medieval science was the exaggerated respect that was paid to the works of the Greeks, and particularly to Plato and Aristotle. The fusion of the number magic of Plato with the quality hierarchy of Aristotle was a multiplication of nonsense from which Islamic science was never able to shake itself free. It is, however, interesting to notice that, though the two great mystifications of early science, astrology and alchemy, were also pursued by the Arabs, the greatest figures of Islamic science such as al-Kindi, Rhazes, and Avicenna explicitly repudiated the extravagant claims of these pseudosciences.

The social position of the scientists in early Islamic culture was not essentially different from what it had been in late classical times. With the coming of the Abbasid dynasty there was a short period between 754 and 861 under the Caliphs – al-Mansur, Haroun-al-Raschid, al-Mamun, and even under the devout al-Mutawahkil – when science was encouraged on a scale unequalled since the early days of the Museum at Alexandria. The Omayyad Caliphs at Cordoba (A.D. 928–1031) and the petty Emirs who succeeded them in Spain and Morocco, were no less

75. Islamic medical men like Rhazes (d. *c.* 924) and al-Kindi (d. *c.* 873) took an interest in other branches of natural knowledge. Much of their learning was derived from Greek and Roman sources, and this illustration of a physician coming to the aid of a man attacked by a snake comes from a thirteenth-century medical manuscript of Galen.

attentive, and even in the decay of Muslim culture ambitious princes such as Saladin, Mahmud of Ghazni, and Ulugh Beg of Samarkand prided themselves on encouraging science. In addition, rich merchants and officials, such as the Persian family of the Barmecides (*c.* 750–803) and the three brothers Musa (*c.* 850), supported scientists and some were themselves interested in science. This secular and commercial background to Islamic science, however, marked it off sharply from that of medieval Christendom, which was almost exclusively clerical (pp. 301 f.). It resembled far more that of the Renaissance. It was this courtly and wealthy patronage that enabled the doctors and astronomers of Islam to carry out their experiments and make their observations. It also protected them, while it lasted, from the active disapproval of religious bigots who suspected that all this philosophy would shake the beliefs of the faithful.

This association of science with kings, wealthy merchants, and nobles was immediately the source of its strength and ultimately of its weakness, since science became, as time went on, completely cut off from the people, who suspected that the learned advisers of the great were up to no good, and this made them an easy prey to religious fanaticism. As long as the cities and trade flourished there was a sufficiently large, cultivated middle class interested in science to ensure discussion and progress. As this broke down, however, the scientists became more and more wandering scholars, dependent on the varying fortunes of local dynasties. Even the greatest of them, Ibn Sina (Avicenna), was never granted any security. He served various Sultans in Persia and Central Asia, sometimes as doctor, sometimes as vizier. At Hamadan he escaped by a feigned illness from mutineers who were demanding his head. Ibn Khaldun (1332–1406), the last of the great Muslim thinkers, was a refugee from Seville forced to take service wherever he could find it. In his time he was to negotiate with Pedro the Cruel in Spain and Tamerlane in Syria, both of whom offered to employ him[3.37] (p. 1044).

THE CHARACTER OF ISLAMIC SCIENCE

The scientists of Islam on the whole accepted and codified the late classical pattern of the sciences. They had little ambition to improve it and none to revolutionize it. As al-Biruni (973–1048) put it, 'We ought to confine ourselves to what the Ancients have dealt with and to perfect what can be perfected.'[3.13.376] Though individuals might specialize, science formed a unity cemented by philosophy. It comprised the twin disciplines of astronomy and medicine, united by a more or less admitted astrology which furnished the link between the outer big world of the heavens –

76. Man the microcosm echoed through his being – mentally and physically – the macrocosm of the universe. Such was the belief of Islamic and medieval philosophers whose concept was of man as the epitome of divine creation and of a universe subservient to him. From Robert Fludd, *Utriusque Cosmi . . . Historia*, Oppenheim, 1617–19.

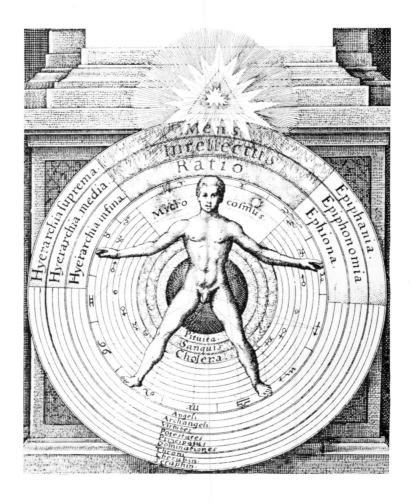

the *macrocosm* – and the inner small world of man – the *microcosm*. Philosophy as such was suspect – it was difficult to reconcile it with the Koran. Pious Muslim scholars certainly attempted to do so but this was frowned on by the orthodox. Al-Ghazzali's (1058–1111) book, *The Destruction of the Philosophers*, was a warning of the futility of this attempt. Despite the spirited answer of Ibn Rushd (1129–98), the much-maligned Averroës, in his *Destruction of the Destruction*, the warning remained effective, and inevitably forced the doctrine of two truths – a higher spiritual and a lower rational truth – which ultimately sterilized both in Islamic countries as surely as it had done among the Greek Christians. The ultimate failure to associate science with the enduring features of Muslim religion was probably a major reason for its withering away in the later centuries of Islam, which became culturally and intellectually static.

During the most flourishing period of Islamic science, in the ninth, tenth, and eleventh centuries, these considerations did not yet weigh heavily. Indeed one may suspect that religion was taken for granted by some of the greatest scientists, and not allowed to interfere with the pursuit of secular knowledge. The unity of science was further ensured by the tradition of encyclopedism, which drove all the great and a number of minor Islamic writers to compose comprehensive treatises like the *Compendium of Astronomy* of al-Fargani (Alfraganus) (d. c. 850) and the great medical collections – the *Howi, Liber Continens* of Rhazes (865–925), the *Canon* of Avicenna, and the *Colliget* of Averroës – which were still being used as textbooks in seventeenth-century Europe.

This comprehensive tendency was all the more valuable because its wider inclusion of the knowledge of other countries gave Islamic science a distinct advantage over that of classical times. Not only were the Arabs able to make use of the Mesopotamian astronomical and mathematical tradition, which had continued unbroken since Babylonian times, but they consciously used the ancient knowledge of India and to a lesser extent that of China.

MATHEMATICS

The central interest in astronomy for its philosophical and astrological implications carried with it a renewed interest in mathematics, as astronomy was almost the only field of mathematical application and encouraged the pursuit of both geometry and computation. Here Islamic mathematicians, owing largely to Babylonian and Indian influence, made their greatest advance. The manipulation of numbers which appeared, with Diophantus, late in Greek mathematics (p. 214) was further developed, helped by the general introduction on a large scale of the Hindu system of numbers, which was already known, though not much used, by the Syrians. This technical device had almost the same effect on arithmetic as the discovery of the alphabet on writing. Before then, arithmetic, other than what can be done on the fingers or the abacus, was a mystery which only the most learned understood. With arabic numbers it was within the reach of any warehouse clerk; they democratized mathematics. The Arabs also incorporated the work of a series of Indian mathematicians on the means of dealing with unknown quantities which we call *algebra*. The word itself comes from the title of al-Khwarizmi's great compendium *Hisab al-Jabr w-al-Muqabalah* or 'restoration and reduction' as a means of solving equations. The Arabs also developed much further another field of great importance both to astronomy and surveying, that of *trigonometry*.

ASTRONOMY

In astronomy the Arabs carried on the Greek tradition, accepting, without searching criticism or any radical advance, the elaborations of Ptolemy (p. 216), whose *Almagest* (Megale syntaxis) they translated. If they did not add to theory, they did keep up without a break the astronomical observations of the Ancients. In particular the observatories of Harran, a city of Chaldean star worshippers, continued well into Abbasid times, protected from Islamic interference by the fiction that they were Sabean 'people of the book'. If a break had been allowed, the Renaissance astronomers would not have had some 900 years of observations behind them, and the crucial discoveries on which modern science rests might have been made far later or not at all.

GEOGRAPHY

Geography remained for Islamic scientists what it had been for the Greeks, a special branch of astronomy. Yet, while they made little theoretical advance, on the practical side they were able to add to the knowledge of the Greeks to such an extent that they laid the foundation of the modern geography of Asia and North Africa. This they owed to the wider range of the Islamic world and the decentralization of its culture – for learned men were to be found from Fez to Samarkand – and to the long journeys that were undertaken by traders and pilgrims to Mecca. The traders penetrated far outside the countries of Islam itself. Learned travellers like al-Masudi (A.D. 900–57) went into Russia and Central Africa and all over India and China, and many of them wrote well-ordered and rational accounts of their journeys, far in advance of the

77. The astrolabe was used by Muslim and medieval navigators for measuring celestial altitudes, for surveying and, indeed, as a calculator for latitude. The manufacture of astrolabes continued well into the eighteenth century.

legends and marvels of the medieval geographers of Europe. Al-Biruni in his great book on *India* gave not only a description of its physical features, but an account of the social system, the religious beliefs, and the scientific attainments of the Hindus in a way that was not to be equalled until the eighteenth century. Geography was not merely descriptive, it was also metrical. Maps and charts were made and astronomical instruments used in navigation. The Caliph al-Mamun (*fl.c.* 830) ordered two measurements of a degree of latitude to be made. In this he had been unwittingly anticipated by I-Hsing in China[3.8] but it was not till the sixteenth century that his performance was to be improved on in Europe by Fernel (p. 404).

ISLAMIC MEDICINE

Islamic medicine, like Islamic astronomy, was a direct continuation of that of the Greeks. Added to it, however, was a knowledge of new diseases and drugs, which was made possible by the wider geographical spread of Islam. The doctors, not only Muslim but also Jewish, studied a great range of diseases and concerned themselves as well with questions of the effects of climate, hygiene, and diet, not neglecting the practical art of cookery. Serving as they did rulers and wealthy merchants, the prestige of doctors was very high as were their intellectual standards. The great Islamic doctors like Rhazes and Avicenna were men of wide knowledge which ranged from astronomy for astrological purposes, through botany to chemistry, for the selection and preparation of drugs. The fact that nearly all Islamic scholars were doctors, and practising doctors at that, had an important and not sufficiently recognized influence on their scientific and philosophic views.

OPTICS

One branch of medicine which was much developed was the study of eye diseases, probably because of their prevalence in desert and tropical countries. The surgical treatment of eye conditions led to a renewed interest in the structure of the eye. This was to give the Arab physicians the first real understanding of dioptrics (p. 218) in the new sense of the passage of light through transparent bodies, and hence to lead to the foundation of modern *optics*. The lens of the eye was to point the way to the use of crystal (*beryllus – brillen*) or glass *lenses* for magnification and reading, particularly by the old. The device of mounting such lenses in frames as spectacles was to come later (p. 320). The *Optical Thesaurus* of Ibn al-Haitham (Alhazen) (*c.* 1038) was the first serious scientific treatment of the subject, on which all medieval optics was based (p. 302). Although improved on, it was not to be superseded till the seventeenth century. In the lens we have the first extension of man's sensory apparatus, to balance that of his motor capacity already achieved through the use of mechanics. It was to be the prototype of the telescopes, microscopes, camera, and other optical instruments of later times. If they had done nothing else, the Islamic doctors in founding optics would have made a decisive contribution to science.

THE BEGINNINGS OF SCIENTIFIC CHEMISTRY

It was in chemistry, however, that the Islamic doctors, perfumers, and metallurgists made their greatest contribution to the general advance of science. Their success in this field was largely due to their having escaped, to a considerable degree, from the class prejudices which kept the Greeks away from the manual arts. Their treatises show a direct acquaintance with laboratory techniques in the handling of drugs, salts, and precious metals. The Arabs were not the first chemists, and worked on the basis of traditions and practices already deeply rooted in Egyptian and Babylonian civilizations and only slightly rationalized by the Greeks. They were also able to draw, to an extent difficult to ascertain, on the extensive chemical knowledge of the Indians and Chinese. [3.8] Chemistry, to a different degree from astronomy and mechanics, depends on widespread experience of large numbers of substances and processes. It can only become a science if these can be brought together and transmitted as a graspable whole, and provided with some general principles. This is what the Arabs did, and what justifies their claim to be the founders of chemistry.

One practical key to chemical advance, the still, had already been discovered in its earlier form of the kerotakis or alembic (ambix) (p. 222), but the Arab chemists much improved it and used it for a large-scale *distillation* of perfume.[2.28] If it had not been for the Koranic prohibition of wine they might have made the next crucial advance and distilled

78. In alchemy the purest substances were sought, for only from them could the mystery of transmutation be achieved. Distillation was one of the time-honoured methods of purification. Drawing prepared from an Arabic alchemical manuscript.

alcohol, but that was apparently left for the Christians. The wealth of new techniques, of which this was only one, was not, as was much of the technique of classical times, left to find its way into craft traditions. It was examined and discussed by the most able doctors and philosophers. Accordingly, it was possible for the first time to approach chemical transformation rationally, though, on account of its objectively greater complexity, never with the same simple analysis that sufficed for mechanics or astronomy.

Instead, chemical ideas originated from the actual method of thinking by analogy, that is essentially biological or sociological.* In chemistry there is a fundamental duality – which we now know is due to shortage or excess of electrons – exemplified by metals and non-metals. There is evidence for tracing the first appreciation of this duality to the Chinese, who already in prehistoric times used red cinnabar as a magic substitute for life blood and had resolved it into its elements, sulphur and mercury. Identifying these with the generalized male and female principles, the Yang and the Yin, themselves of totemistic origin, the Taoist sect developed a system of alchemy from which it is probable that first Indian and then Arabic alchemy was derived. It was originally not so much a method of preparing gold but the elixir of life.[3.36]

The Arabs took up this mercury–sulphur theory and extended it.[3.58;3.60] It was to be the germ of the spagyric theory of Paracelsus (p. 398) and, through him, first of phlogistic and then of modern chemistry. The earliest writings seem to have been lost or possibly incorporated in the pseudo-Aristotelian doctrine of dry and wet earthy exhalations, used to explain the origin of minerals. Similar ideas have been attributed to Jabir (Geber), who is supposed to have flourished in the eighth century and to be the father of Arabic chemistry. However that may be, there is certainly found in the works of al-Razi (Rhazes), the greatest of Arab physicians, an extensive compendium of chemical operations and substances. The future of chemistry was indeed to depend on the first full-scale production in localized chemical industries in Islamic countries of such commodities as soda, alum, copperas (iron sulphate), nitre, and other salts which could be exported and used, particularly in the textile industry, all over the world.[3.25; 3.58]

THE LEGACY OF ISLAMIC SCIENCE

This bare outline can do but scant justice to the extent and weight of the Islamic contribution to science. Though in its central themes it is evidently a continuation of Greek science, the latter was both revivified and extended. By their renewed activity as much as by their search for earlier and better authorities, the Islamic scholars rescued Greek science from the decadent state that it had fallen into under the late Roman Empire. They created a live and growing science, even if at no point does it rise to the heights of the speculations of the Ionic Nature philosophers or equal the geometrical imagination of the Alexandrian school. By drawing on the experience of non-Hellenic countries, Persia, India, and China, they were, however, able to extend the narrow basis of Greek mathematical, astronomical, and medical science, to initiate the techniques of algebra and trigonometry, and to lay the foundations of optics. The crucial extension of Islamic science was to be in chemistry or alchemy, where they transformed old theories and added new experiment to create a new discipline and tradition of science. This tradition was often qualitative and mystical in character, but for that very reason was to be, for many centuries, an invaluable counterweight to the over-rational and mathematical, astronomical–medical tradition of the Greeks.

5.7 The Decay of Islamic Culture

After the eleventh century, although there was no spectacular collapse, it is evident that the best days of Islamic science were over. There were still brilliant individual scientists. One of the greatest, Averroes, dates from the twelfth century, and Ibn Khaldun comes as late as the fourteenth, but they are no longer part of a widely based and living movement. The failure of science is here only one symptom of a general political and economic decay of Islam in its original form. Essentially it was the delayed effect of the same social forces that had brought about the decay of classical culture. Neither in Islam nor in the surviving Eastern Roman Empire of Byzantium could the same inequalities in wealth fail in the long run to lead to economic breakdown. The Arabs, when they took over the Asiatic provinces of the Empire, inherited its problems as well as its wealth. The subjection of peasants and craftsmen destroyed the market for an effective industry. This result could only be postponed by using up the considerable resources accumulated in the Byzantine Empire, and by opening new fields for commercial exploitation in Russia, Central Asia, and Africa.

In the end both the Byzantine and Islamic Empires were unable to maintain the organization necessary to control an extensive State. By the tenth century they both began to break up internally, and to grow

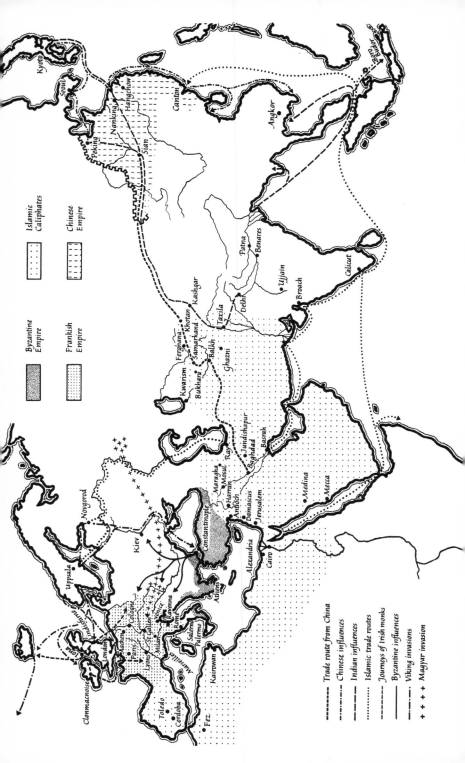

more dependent for military, and then for economic, purposes on local efforts. By the time of the Crusades both had developed into a local feudalism which was inferior militarily, and culturally was no longer markedly superior, to that of the West. Further, the Eastern feudalism lacked, as we shall see, the economic resources and the cultural hopefulness of the new feudalism of the West. It lacked especially the widespread basis that was provided by the manorial village, with its living tradition of the old tribal collective.

The breakdown of Islamic civilization was undoubtedly accelerated by the arrival of new waves of barbarians from the Steppe lands. The Turks and Mongols by themselves, however, would never have been able to overrun the Islamic lands and effectively sterilize their culture by the thirteenth century, had these been in a flourishing economic state. As it was, the irrigation agriculture of Mesopotamia was largely ruined by a combination of native misrule and Mongol incursions which prevented the maintenance of the canals.* That invasion alone is not a sufficient explanation is shown by the contemporary decline of Egypt and north Africa, into which the Mongols never penetrated, and the fact that similar incursions into the fundamentally more stable economics of China and India had no effect on their economies and little on their cultures.

Islam was to survive and survives to this day as a religion and a civilization, but it was not to regain the same scientific impetus that marked its first flowering. The equilibrium reached in the Mongol and Turkish States that succeeded the original Arab empires was one in which science stayed substantially frozen at the stage it had reached in the eleventh century. The ostensible reason for this was the rise of the clerical faction which actively discouraged philosophy and science. But

Map 2

The World of the Transition to Feudalism, A.D. 550-1150

This map brings out the relations of the different centres of civilization and the approximate extent of the empires as they were in the mid-eighth century. The Byzantine and Irish monastic influences came earlier in the eighth century; the Viking and Magyar raids later towards the end of the ninth. The towns marked, particularly in Central Asia, are centres of trade and science.

this, had there been any real need for science, would in itself have been no more effective than it was in Renaissance Europe. In the East, once the earlier stimulus to economic progress failed, the intellectual stimulus also vanished. Both might have revived later, but by the time they showed signs of this, as in India under the Moguls, their development was cut short by the superior commercial and military achievements of early European capitalism.

The fruits of Islamic science were, however, not wasted, though they were not to be enjoyed in the lands that cultivated them. To an extent far greater than that of the transmission of Greek science, the whole apparatus of Islamic science, data, experiments, theories, and methods were handed on directly to the new, growing science of feudal Christendom. Indeed, if this were a history of science instead of one of its influence, it would have been more logical to treat together as one chapter of intellectual advance the whole period from the seventh to the fourteenth century, hardly distinguishing the languages – Syriac, Persian, Hindi, Arabic, or Latin – in which the books were written. The difference between the new science of the sixteenth century and that of the thirteenth in Europe is far greater than that between Arabic and Latin science in the twelfth century. Both the glory and the limitations of Islamic and Christian science in the Middle Ages stem from the same root, their association with the political and economic basis of feudalism; but the demonstration of this must await the next chapter.

Chapter 6

Medieval Science and Technique

6.1 The Dark Ages in Western Europe

While a brilliant cultural development was taking place in the Eastern Empires and in Islam, most of Europe was still suffering from the confusion left by the collapse of the Roman Empire and by the barbarian invasions. Between the fifth and the ninth centuries towns decayed everywhere. In Britain, where they were alien foundations, they completely disappeared; in Italy, where they had 1,000 years behind them, they survived, though half ruined and deserted. The first barbarian rulers – Franks and Goths in the West, Slavs in the East – maintained a shadow of the Imperial system, including trade on a considerable scale in luxury goods and slaves. Classical culture gradually died out, leaving such living relics as the swan song of Boëthius. The new Christian culture, preserving the scriptures and fragments of Latin and Greek literature, was spread from outlying centres such as Iona[3.4] or Kiev.[3.32] Only in Constantinople was a Christianized Empire, more Greek than Roman, able to maintain itself and to guard the classical heritage.

The western kingdoms, despite their unification under Charlemagne, were unable to maintain a State organization on the Roman model against the treble attack of the Normans, Magyars, and Saracens. Nevertheless they were not overwhelmed, and emerged after a few years vigorous but fragmented. Their successful resistance was achieved on a basis of local defence and local self-sufficiency – the feudal system. Once this was well established, as it was from the year 1000, recovery was rapid. The very factors which had held up the early development of western Europe – its forests and heavy soils – made its advance more rapid when it came. From the tenth century onwards the intrinsic economic advantages of Europe began to tell. They were primarily agricultural, based on the suitability of western European climate and soils for dry cultivation once technical difficulties of cutting down woods Islamic East, on the other hand, was for the most part an arid region. As and ploughing heavy soils could be overcome (pp. 234, 314). The such, it was liable to increased desiccation and erosion, and this became catastrophic when it was combined with the decay of the governmental organization which alone could maintain irrigation systems and keep the ravages of faulty agriculture in check.

No such requirement for extensive organization existed for western Europe; only local and not national effort was required. Even starting from a stage of extreme disorganization, its economy could rebuild itself village by village. Slowly but irresistibly a new civilization, which was soon to surpass its forerunners, arose on a solid basis of abundant, fertile, and well-worked land. Nevertheless only the western and northern parts of Europe were able for long to make use of these advantages. They were saved by their remoteness, and even more by their forests, from the last incursions of the Asiatic pastoral peoples. In the thirteenth century the Tartars overwhelmed the highly civilized State of Kiev. This Byzantine equivalent of Charlemagne's Frankish Holy Roman Empire was not entirely wiped out, but had to be recreated from its offshoots in the northern forests. As a consequence the Russian State came into action, as Great Muscovy, some centuries later than the west of Europe. In the fourteenth and fifteenth centuries the same fate fell on south-eastern Europe when the south Slav kingdoms and finally Byzantium itself were overrun by the Turks.

The world of medieval Christendom was thus a very limited one. Its central spine ran from Italy through eastern France to England; to the east it included only the Rhineland and the Low Countries; to the west, Gascony and Catalonia. Even in this area the most characteristic

79. Peasant harrowing: English, fourteenth century. From the *Luttrell Psalter* in the British Museum (Additional MS 42130 [XIV cent.]).

developments were more limited still, centring on the rich, well-watered agricultural plains of Flanders, Normandy, Champagne, and the Paris basin, and on the southern counties of England. It was in the land of the Franks, in the very Île de France of which Paris is the centre, that the economic forms, the architecture, and the intellectual developments of medieval scholarship first came to flower. The other great cultural centre, that of Italy and particularly of Lombardy and Tuscany, was too impregnated with the influence of the classical world to produce such distinctive contributions. Its turn was to come in the later Middle Ages and the Renaissance (Map 3, p. 338).*

6.2 The Feudal System

In contrast to the slave economy of classical times that preceded it and that of capitalist economy that followed it, the economy of the whole period from the fifth to the seventeenth century may be taken as feudal (p. 253). Nevertheless it is only in Europe from the eleventh to the fourteenth centuries that the *feudal system* appears fully developed, complete with its political and religious hierarchies and with the corresponding art and knowledge.[3.30]†

The economic basis of the feudal system was the land. It was marked by its dependence on local agricultural production, largely consumed on the spot, and on a scattered handicraft industry. The economic unit in the feudal system was the *village*. There, some scores of men and women, mostly kinsfolk, shared out the land and work, holding most in common. They were not far removed in sentiment and sometimes even in ancestry from the old clan groupings. They carried out a simple rotation of crops, usually, in northern lands, with three fields divided into individual plough strips and some woods and pasture. On the peasants was superimposed a hierarchy of lords, lay or clerical, and their overlords, bishops and kings, under the nominal headship of emperor and pope. Each lord might hold one or more villages, or land in several villages, where his serfs were obliged to work to keep him as well as themselves. It is this obligation of feudal service, that is of work exacted by force or by custom backed by force, that distinguishes feudal exploitation from the wage-labour system of capitalism. It is the imposition of this obligation on peasants with secure tenure cultivating their own land that distinguishes it from the chattel slavery of classical times.

In theory, feudal obligations were not entirely one-sided. In return for the service of his peasants the lord was supposed to give them protection, but this should be understood rather in the gangster sense.* For the commonest danger against which he had to protect them was the attacks of other lords. The whole duty of a noble lord was to fight for his overlord when called on, though he might fight against him when he felt like it. For the rest, he could eat and hunt. The whole duty of the spiritual lord was to pray, but he usually managed to consume as much provender to support him in this as his lay brother. The higher nobility, lay and spiritual, together with their retainers, had for lack of adequate transport virtually to eat their way round their scattered manors. Even the king could never afford to stay in one place for long, but had to travel round with his court like a circus.[3.59] The nobility and clergy of the feudal system were little more than parasitic on the village economy. This parasitism, however, was thorough and intelligent. The bailiffs of the manor, lay or clerical, had learned well how to extract the last ounce of service and dues from the serfs.[3.59]

The fact that it was possible, without large-scale trade or organization, to maintain a parasitic class which with their unproductive retainers amounted to some ten per cent of the population, shows that the economy of the feudal village was far from primitive. Though in its social form it represents a return to a pre-classical village economy, it was a return on a higher technical level, with widespread use of iron, better ploughs, better harness, better looms, and the use of labour-saving devices such as the mill. The technical advances of classical times, which were concentrated in the cities and where production on the slave plantation villas was for the benefit of a plutocracy of traders and landowners, were, in feudal times, spread widely over the countryside, giving everywhere a local surplus. The feudal system was therefore, technically as well as socially, a far more secure base for further progress than was classical plutocracy.

At the same time, it was too locally subdivided and lacking in concentration to achieve this progress rapidly from its own internal initiative. What it could and did do, particularly from the eleventh to the thirteenth centuries, was to spread over the untilled and waste parts of Europe. This spread of land cultivation represented the only way in which feudal economy could develop without losing its character. It was pushed forward by nobles and churchmen alike, eager to enlarge their estates and power, and it was often supported by serfs as well because they could bargain for better conditions in the new lands. By the end of the thirteenth century this expansion overshot itself, and led to a serious economic crisis from which feudalism never really recovered.

Meanwhile, however, other economic forms were growing up inside the feudal system, based on a trading and urban manufacturing economy. These, by breaking down the local self-sufficiency of feudal economy, were ultimately to destroy it; but at first they could be assimilated into the feudal system, which was to continue for another two centuries in Britain and Flanders, and for longer still in the rest of Europe. Feudal economy itself was largely a product of the disorganization produced by the collapse of the classical economy, and the barbarian invasions and disturbances this provoked. Once conditions settled down and warfare became merely occasional, tendencies to forms of organization not so directly based on the land reasserted themselves.

THE MEDIEVAL TOWNS

Starting in the Mediterranean area, in south Italy, Provence, and Catalonia, where they had suffered least in the Dark Ages, and soon after

80. A painting of a city thought to be by the fourteenth-century artist Ambrogio Lorenzetti (?1323–48?). It shows beautifully the medieval walled city as it began to develop after the Dark Ages.

81. Wind power was one of the important sources of mechanical energy until the advent of the steam engine in the eighteenth century. From the 1620 German edition of *Le Diverse et Artificiose Machine* by Agostino Ramelli (*c.* 1530–90).

in the Rhineland, the Low Countries, and Lombardy, where the agricultural surplus was greatest, towns began to grow again.[3.48] By the eleventh century towns were well established in these areas; by the twelfth they were also growing in northern France, England, and Germany east of the Rhine. As they grew they strove to emancipate themselves from the restrictions of Church and feudal institutions. In Germany and Italy, where central government was weakest, they became virtually independent city states; in France and England they remained subordinate to royal, though not to feudal, power. These towns lived by exchanging new manufactured goods, made by guilds of handicraftsmen within their walls, for the surplus products of feudal economy. The towns contained at first a negligible proportion of the population; even at the end of the Middle Ages in more urbanized countries such as Italy and Flanders they represented probably not more than five per cent. Nevertheless, their establishment was of crucial importance because it was from them that ultimately was to come the bourgeois (burgess) class that was to found capitalism. The same urban movement was also to be the focus of a new utilitarian science, radically different from that of the Ancients.

Throughout most of the Middle Ages, however, the towns had not this revolutionary role. Once they had achieved their necessary liberties they fitted very well into the essentially rural feudal economy. This economy, however, was by no means a stable one. In its first phase, as already indicated, the main emphasis was on the establishment and extension of feudal order.[3.48] After the thirteenth century that order itself was beginning to break down not merely in Italy, where it had been least securely established, but at its centre in the Low Countries, England, and northern France. That breakdown was on the whole a progressive and not a degenerative one. It was marked by an increased production, not only of food but also of textiles, accompanied by a differentiation of peasantry in which the richer, at least, became emancipated from feudal service. Commodity production for the market took the place of subsistence economy, with a resulting enhancement of the importance of trade and towns. These were the conditions that gave still further impetus to the technical changes in manufacture and transport that were to lead to the new age of capitalism.

The impetus to technical innovation had, however, existed from the beginning of the Middle Ages, particularly in the better utilization of land and the increased use of machinery. It was here that the medieval peasant and workman could profit by the legacy of classical techniques and by the addition that the Arabs had made to them. What had been lost was largely, as already indicated, the arts of luxury and of large city organization. Aqueducts and baths could be done without, but mills and smithies remained. Agriculture and the practical arts were further improved, as we shall see, by borrowings from the East and by indigenous inventions. This improvement took the direction of a substitution of mechanical for human action; of animal and water-power for man power. There was nothing, it is true, that the medieval craftsman could do that could not have been done by the Greeks or Romans, but they lacked the compelling incentive, the need to do more work with fewer men.

For most of the Middle Ages there was a chronic labour shortage. It was not only that there was no longer the expendable labour force of slaves that had held up technical advance in classical times. There was also the drive for extension of cultivation that stemmed from the nature of the feudal system. The nobles needed more and more land, but land was useless without peasants, and there were never enough of them, especially at harvest time. Of course, peasants could be made to work harder and hand over more of the produce to the lord, but there was a limit to this, forcibly demonstrated in peasant revolts. Hence the search, first by enterprising feudal lords and ecclesiastics, then by wealthy merchants, for alternative methods of enrichment – for mills, for textile factories, for mines, and for foreign trade. Technical progress was slow, held up by the vested interests of nobles and guildsmen, but it could not be arrested, and its consequences in the end were to sap the foundations of the feudal system and the medieval world order which was its intellectual expression.

6.3 The Church in the Middle Ages

The feudal system furnished the economic basis throughout the Middle Ages; its intellectual and administrative expression was provided by the Church. It was the unity and order of the Church that counteracted the anarchic tendencies of the nobles and provided for all Christendom a common basis of authority. Though on particular issues there was often a conflict of power between emperor and pope, king and bishop, both sides recognized the need of the other in the maintenance of society. The Church did not stand out against the feudal system, it was an essential part of it, and indeed one could not be changed without the other, as the Reformation was to show.

In the age of transition before the tenth century, the Church in the West was most concerned with the mere business of cultural survival. It was the one rallying ground of ancient civilization against successive waves of barbarians, Goths, Vandals, Franks, Saxons, and Lombards, who as they came into the pale of the Roman Empire had to be won to Christianity. Later the effort of conversion spread further, to the Norsemen and the Magyars. In all cases it imposed its rule in the first place as the heir to the greatness of the Empire, appealing to the ambition of barbarian chiefs and to the credulity and love of wonder of their households. In the process it was inevitable that the Church itself became barbarized; though it clung to the impressive externals of religion, the rituals, vestments, relics, and miracles, it lost much of its early intellectual content. What was saved was through the efforts of the remote early missions from Ireland and Northumbria, where monks such as Bede (673–735) and Erigena (*c*. 800–*c*. 877) preserved something of classical scholarship and philosophy.[3,4]

The first general movement of intellectual recovery in Europe was that of Charles the Great, who, though himself illiterate, introduced palace schools in the ninth century; but that was set back by new invasions of Norsemen, Magyars, and Saracens. It was only in the tenth century with the monastic reformation, starting at Cluny in Burgundy, that the Church seriously began to build an organization which could control the lives and thoughts of all the people of Christendom from king to serf. This organization was itself feudal, and indeed doubly so, for not only were the hierarchy of secular churchmen, popes, archbishops, bishops, and priests, all feudal landowners, but also the regular clergy, the monks, actually opened up land on their own account in their abbeys and were the spearhead of feudal expansion.

All through the early Middle Ages, at least up to the beginning of the thirteenth century, even in Italy, the Church had in its priests and monks a practical monopoly of learning and even of literacy. Feudal administration had to pass through clerical hands, as the word clerk bears witness today. This monopoly was to give medieval thought a degree of unity, but it was also seriously to limit its scope. Neither Greek not Islamic thought had been so confined to a single order of men (p. 272).

The professed attitude of the medieval Church to human affairs had been set in the dark days of the decay of the Roman Empire. It was that life in this world was a mere preparation for an eternal life in hell or heaven, an attitude which only gradually weakened with the undeniable improvement of human conditions, but was not to be blown away till the Renaissance. In practice, however, the Church took a shrewd interest in the affairs of this world, and was deeply involved in the maintenance of the feudal order.

THE COMING OF THE FRIARS

This concern with an essentially rural economy set the Church, from the twelfth century onwards, in opposition to the interests of the secular society of merchants and artisans of the new towns. These expressed their dissatisfactions in heresies, usually of a Manichean and mystical kind, maintaining that man could approach God without the mediation of a host of greedy and ill-living clerics. Such heresies could be, for a while, put down by the sword, as in the great Crusade against the Albigenses in 1209; but by the mid thirteenth century a more satisfactory solution was found. The Church secured a new arm in the licensed beggars and preachers – the Franciscan and Dominican friars – who had come into existence partly as an expression of, and partly as a reaction to, the changed conditions.

St Francis of Assisi (1182–1226) reflected in his life and preaching the revolt of the poorer townsmen against worldliness and excessive wealth. His message was popular, dangerously so, and all of papal diplomacy was needed to keep it from breaking out in heresy and civil strife. Such difficulties are still found today in the handling of 'worker priests' in France. Even after the resistance of the 'spiritual' Franciscans had been broken in 1312, their doctrine continued to work through Occam (d. *c.* 1349) and Wycliffe (*c.* 1324–84) and paved the way for the Reformation.

The friar preachers of St Dominic were, on the other hand, deliberately reactionary from the start. Their ostensible aim was to use persuasion to check the spread of heresy. The townsmen were becoming intelligent, even learned, and the greatest weight of orthodox learning had to be massed against them. Hence the philosophic labours of St Albert (1193–1280) and St Thomas Aquinas (*c.* 1227–74); hence also their instinctive sympathy with Aristotle, the great defender of order. How effective this persuasion was, as compared with the more brutal efforts of the Crusades and the Inquisition, it is difficult to tell, but heresy was kept down for some 300 years.

Nevertheless, despite the efforts of the friars, the last two centuries of the Middle Ages were to witness a definite weakening of the Church under the influence of the rising towns and the growing strength of the kings, who were increasingly allying themselves with the towns against the country nobility. The papacy was forcibly moved to Avignon in 1309 and the Church was split between two or three popes from 1378 to 1418.

In healing this breach new authority was vested in general councils. Even these could not keep order, and though they could burn Huss in 1415, his followers defied them and maintained an independent national State in Bohemia until 1526. The Church, however, was only weakened as an organization; it had so impressed its stamp on intellectual and social thought that the disputes in politics and science of the next few hundred years were to be carried out mainly in terms of religion.

6.4 The Scholastics and the Universities

The revival of western Christendom which began in the tenth century required an intellectual basis wider than was provided by the meagre salvage of classical lore, even where transmitted by such able thinkers as Bede and Erigena. The clergy had to be trained to think and write; the claims of the Church, spiritual and temporal, had to be asserted and defended. At first this need was met by the setting up of cathedral schools such as those of Chartres and Reims. By the twelfth century these had swelled to become *universities* with set courses teaching the seven liberal arts, philosophy, and, most important of all, theology. The first and most famous of these, the university of Paris, was not so much founded as recognized by 1160. The idea of a university – *studium generale* -- where all subjects could be studied together was not entirely a new one. In antiquity there had been the schools of Athens and the Museum of Alexandria; the Muslims had had their Mosque schools, Madrasseh, for centuries, where philosophy as well as religion had been taught; and already, since the eleventh century A.D., a medical school had been in existence in Salerno. Though the new medieval universities were to borrow from all of these, they were more general and systematic in their teaching, and early acquired a special place in the world of Christendom as repositories of learning. Bologna was founded as early, if not earlier, than Paris; Oxford, practically a branch house of Paris, in 1167; Cambridge in 1209. Then came Padua, 1222, Naples, 1224, Salamanca, 1227, Prague, 1347, Cracow, 1364, Vienna, 1367, and St Andrews, 1410.

From their very foundations the universities were, and remained until relatively recent times, mainly institutions for training the clergy. This emphasis mattered little at a time when the clergy had the monopoly of literate occupations and were responsible for all administration. What was important then was that they should be educated at all, and particularly

82. Late medieval teaching: a woodcut showing the fixed attitude of mind in the teaching of science still prevalent in the fifteenth century. The professor is enthroned in a chair, the teaching of Galen with him. The demonstrator (below) dissected the corpse and demonstrated the veracity of Galen's beliefs; there was no attempt at independent investigation. From the title-page of *Anathomia*, by Mondino de'Luzzi (Mundinus), Leipzig, 1493.

that they should absorb something of the ideas of the classical world. The teaching was by means of lectures and disputations, for books were scarce. This was still the method when faculties of medicine were added. The curriculum was fixed on the basis of the seven liberal arts, a summary, excessively simplified, of classical learning. The first three 'trivial' subjects were grammar, rhetoric, and logic, aimed at teaching the student to talk and write sense – naturally in Latin. Then followed the 'quadrivium' of arithmetic, geometry, astronomy, and music. Only after this study could philosophy and theology be approached. It is significant to note that the basic study was not only secular but scientific; in this it followed the Islamic model. Law and medicine were catered for in other faculties, but neither history nor literature found any place. It was this omission which was to occasion in the Renaissance the humanist reaction against the whole scholastic system (pp. 382 f).

In practice the science taught amounted to very little.[3.5] Arithmetic was numeration; geometry the first three books of Euclid; astronomy hardly got further than the calendar and how to compute the date of Easter; and the physics and music were very remote and Platonic. There was little contact, and little desire for it, with the world of Nature or the practical arts, but at least a love of knowledge and an interest in argument was fostered. In the latter Middle Ages the universities, with few exceptions, such as Padua (p. 392), had come to be guardians of established knowledge and barriers to any cultural advance, but in their early days they were the focus of intellectual life in Europe.

THE IMPACT OF ARAB AND GREEK KNOWLEDGE

It was into this world of restricted and avid intellectual activity that there came the impact of Arab scholarship, carrying with it a far richer draught of classical knowledge than had even been preserved in the West. Beginning with a few works in the eleventh century it came in a full flood in the twelfth, when the bulk of Arab and Greek classics were translated into Latin, mostly from the Arabic,[3.5] but some directly from the Greek. Most of the translation was done in Spain, some in Sicily. The Crusades

had a negligible influence on the spread of culture. This cultural transmission was of a character entirely different from those of earlier times
except, perhaps, of that between Indian and Islamic science. For here,
instead of the passing on of a practically defunct tradition to a new and
vigorous culture, there was a handing over of the fruits of a culture
hardly past its full vigour. At first sight there might have seemed to be
enormous difficulties in the transmission of ideas expressed in a radically
different language, and coming from people with religious beliefs not
only foreign but actively hostile. These obstacles, however, proved to be
superficial compared to the underlying similarity of the culture transmitted by the Arabs to that already held by the Latins. They were, in fact,
only receiving more amply and from closer to its source the Hellenistic
culture that was already the basis of their own. Both contained the same

83. The impact of Greek and Arabic texts on western Christendom was immense.
Until the invention in Europe of printing by movable type at the close of the fourteenth century, all treatises had to be copied by hand. The process was slow and
laborious, though the results were often of great beauty.

substratum of Platonic and Neoplatonic thought. The words were unfamiliar but the meanings were not.

Not only that, but the very religion of Islam had been faced with the same intellectual problems – of the creation of the universe; of the reconciliation between faith and reason; of the literal inspiration or the eternal existence of the Koran; of the validity of mystical experience – that were to perplex the Christians. Duns Scotus and Thomas Aquinas were to continue the dispute already opened between al-Ghazzali and Averroes (p. 275). In terms of science alone, it would be logical to treat the period from the ninth to the fourteenth century as a unitary Arabic–Latin effort to reconcile religion and philosophy and complete the classical world-picture. But this would be to ignore geographical and economic differences that were to bring about a decisive divergence in the consequences of that enterprise. For while in Islamic countries a compromise was reached which sterilized the advance of science, in Christian hands the dispute went on until, under the impact of economic changes, the whole Greek world-picture was destroyed and replaced by another.

FAITH AND REASON

Already in the eleventh century, before the full impact of Arab learning was felt, the disputes of the schools had been turning to this central problem of providing a basis for faith in reason or, more narrowly, of reconciling the scriptures and the fathers with the logic of the Greeks. At first this seemed easy enough: St Anselm (1033–1109) proved the existence of God from that of the idea of perfection. The details of a rational religion, however, were more difficult to fill in. Abelard (1079–1142) indeed presented, in his *Sic et Non*, respectable citations from the fathers expressing opposite opinions on almost every vital question. It seemed at first that the recovery of the major works of Aristotle in the twelfth century would provide sufficient guidance to solve these problems. Indeed his legendary reputation was more than justified when it became possible to appreciate the extent of his knowledge and the rigour of his logic. Moreover, as we have seen (p. 200), the essentially conservative doctrines of Aristotle had been originally made to fit a static, class-divided society. It only needed certain alterations to adapt them to a Christian, feudal, rather than a pagan, slave economy.

The first steps had already been taken by Averroes (p. 275), revered throughout the Middle Ages as the great commentator, but he had too much respect for Aristotle for his version to be readily adaptable to the Christian revelation.[3.29; 3.55] That task was achieved by the Dominican friar, St Thomas Aquinas. His great *Summa Theologica* provides an explanation of the universe of Nature and of man as a framework of the far more important drama of divine governance and human salvation. The whole is arranged in admirable system, with citations for and against every point discussed, together with an argument which always leads to the orthodox solution. Faith is always superior to reason, in the sense that there are things that reason alone could never discover, but equally revelation and reason can never be in conflict. The answers being known in advance, the Saint's arguments often have the air of special pleading. Nevertheless they have never been improved on, and form the basis of Catholic doctrine to this day (*n*. p. 207).

Given the limitation of the time, St Thomas's performance was a remarkable feat of system and ingenuity, for it is more than a mere adaptation of Aristotle: it includes the use of the Aristotelian method to deal with situations of feudal society, which the Greeks could never have come across. Nevertheless it marks no original advance in thought, and to take it today as a philosophic basis is a confession of the intellectual bankruptcy of the neo-Thomist supporters of reaction.

St Thomas, indeed, was too able. Not only did he reconcile with reason the fragmentary and often contradictory doctrines of early Christianity but he used the Neoplatonic forgery, the *Celestial Hierarchy* of the so-called Dionysius the Areopagite – which, to be fair, was taken as gospel by nearly all medieval thinkers – as the major basis for his world order, which is accordingly no more Christian than it is scientific (p. 308).

Some recent historians, impressed with the fact that modern science arose out of medieval scholasticism, have praised the quality of argumentation that enabled the schoolmen to do so. Now in the first place it was not the schoolmen who created modern science, but men like Leonardo, Bacon, and Galileo, who violently repudiated their aims and methods (pp. 439 f.). Further, the history of the scientific revolution shows that the removal of the accretion of nonsense of all ages was far the most difficult and tedious task in the foundation of science. When we realize that it took the best part of 1,000 years to carry out the amount of thinking that, without these obstructions, could have been packed into 200, we may feel less inclined to revere those who established the doctrines that so effectively held back the advance of science.

THE NOMINALIST OPPOSITION

The labours of St Thomas were less kindly received in their times than they were to be long after. Even before the impact of Arab knowledge, there had been an opposition to the highly general method of argumentation based on the *reality* of Platonic ideas or Aristotelian substantive forms. The arguments of Roscellinus (*c.* 1050–*c.* 1122), the first of the *nominalists*, as opposed to the realists, were reinforced despite St Thomas by the Franciscan Duns Scotus (*c.* 1266–1308). The nominalists in effect, by asserting the importance of individuality and maintaining that things came before names or ideas, effectively rejected the whole rational theological scheme. Because they were also good Christians this did not lead them into scepticism, nor for the most part into a direct study of Nature, but rather, like al-Ghazzali, into an assertion of blind faith mystically held and so superior that human reason cannot hope to grasp it. Nevertheless, because they needed to dispute with realists, they had to develop their reasoning in a critical sense, and thus provided arguments that were to be of use in the later revival of natural science. William of Occam's famous razor 'Entities should not be multiplied without reason' or more authentically 'It is vain to do with more what can be done with fewer' has served to remove a lot of nonsense from scientific theory. Later still the school of Buridan (*c.* 1297–1358) and Oresme (1320–82) in Paris used Occam's methods to criticize Aristotle's doctrine of motion, and thus prepared the way for Galileo's reformation of dynamics (pp. 427 f.).[3.2; 3.5] In chemistry, where reason for long had a most tenuous hold, the alchemical approach also found support from the mystically minded. Raymond Lull (*c.* 1235–1315) of Majorca, who was the major source for the introduction of Islamic Sufic mysticism into Christendom, was, or was reputed to be, one of the founders of the chemical tradition which, as will be shown (p. 398), was to run through Paracelsus and van Helmont to the chemistry of the present day.

6.5 Medieval Science

This long theological, philosophic preamble to medieval science is necessary because what little scientific investigation there was in that age was undertaken almost exclusively for religious ends and by clerics – priests, monks, or friars. In this it is in marked contrast to the conditions of Islamic science, where few of the scientists had any religious calling and most had frankly utilitarian ends (pp. 271 f.).

The present fashion of extolling the science of the Middle Ages to the detriment of that of the Renaissance is a particularly silly one. Inaccurate in fact, it is especially unfair to the medieval cleric and scholar, giving him credit where he did not seek it and obscuring his real contribution. Even Roger Bacon (*c.* 1235–1315) in his ill-tempered and perverse denunciations of his contemporaries – he treats the great St Albert and St Thomas as 'ignorant boys' – would never have questioned that the main end of science was the buttressing of revelation.[3.27] His only difference from them is that he sought his confirmation in experience instead of reason. The men of the Middle Ages were perfectly competent in reasoning and in the design and carrying out of experiments. These experiments, however, remained isolated and, like those of the Greeks and Arabs, were essentially demonstrations leading to no decisive advance. Much as they deserve credit for their achievements, the handful of medieval experimenters did not make much use of these methods to investigate Nature and less to control it. They had no incentive to do so and plenty of reasons to dissuade them. Being churchmen they had many other preoccupations: Gerbert (*c.* 930–1003), the first of the western scientists, became a pope; Robert Grosseteste (*c.* 1168–1253), the ablest of them, was a bishop and a chancellor of Oxford University; St Albert the Great was a provincial of the Dominican order responsible for the whole of Germany, so was Dietrich of Freiburg (*fl.* 1300), the best experimenter. Even the most daring thinker of the late Middle Ages, Nicholas of Cusa (1401–64), was drawn into papal propaganda and ended as bishop of Brixen. Anything they did in science was spare-time work.

The exceptions, Roger Bacon and the mysterious Peter the Pilgrim, prove the rule. Roger Bacon spent a large fortune on scientific researches and despite the Pope's blessing was put in prison for his pains. Peter the Pilgrim was a pioneer in the experimental study of magnetism, on which he published one short letter (pp. 317 f.). According to his admirer Roger Bacon, 'He does not care for speeches and battles of words but pursues the works of wisdom and finds peace in them' (p. 310).

The sum total of the medieval achievement in the natural sciences can be put down as a few notes on natural history and minerals by St Albert, an important treatise on sporting birds by the Emperor Frederick II, some improvements in Alhazen's optics by Dietrich of Freiburg and Witelo, including an account of the rainbow that was not to be bettered till Newton, and some not very original criticisms of Aristotle's theory of motion by Buridan and Oresme.[3.5] On the strength of this it is now asserted that the scientific revolution should date from the thirteenth century and that St Albert, somewhat belatedly canonized in 1931, has the right to be patron saint of science.

MATHEMATICS AND ASTRONOMY

In mathematics and astronomy, though the showing is better, it is essentially the same story. Fibonacci (*fl.* 1202), Leonard of Pisa, introduced Arabic algebra and Indian numerals into Christendom. He was a considerable mathematician himself but left no school, and mathematics made no serious advance till Renaissance times. In mechanics, Jordanus Nemorarius (d. *c.* 1237), in a rather simple account of the theory of the lever, advanced the principle of the equality of work done by a machine to that impressed on it, but this had and could have no effect on actual mechanics given the state of technique of the time.

In astronomy Ptolemy's *Almagest* was translated from the Arabic by Gerard of Cremona in 1175. Its study, together with the tables of up-to-date observations composed on the basis of earlier Arabic observations at the order of King Alfonso the Wise in the thirteenth

84. In medieval times there was a return to the concept of a flat Earth and a dogmatism about the crystalline celestial spheres, here epitomized in a woodcut showing the machinery responsible for their motion discovered by an inquirer who has broken through the outer sphere of the stars. Sixteenth century.

century, made possible the continuation of Hellenistic astronomy in Christendom. There, as in Islam, it was of use mainly for calendaric and astrological purposes. It is worth noting that in observational astronomy, the only science where accurate observation, calculation, and prediction were necessary, Islamic predominance lasted longer than in any other branch of science. The Ilkhanic tables of Maragha (*c.* 1260) and those of Ulugh Beg (1394–1449) were the best available until the Renaissance. The medieval astronomers showed themselves capable of making some improvements in detail in astronomical calculations, particularly the

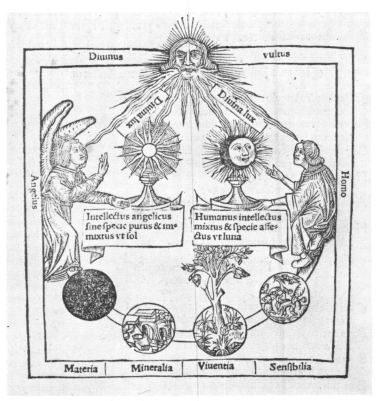

85. The medieval concept of the great chain of being. The divine light illumines both angel and man, who are connected by the kingdoms of unformed matter, minerals, plants and sentient creatures. From *De Intellecto* by Charles Bouelles Bovillus (*c.* 1470–*c.* 1550).

school at Merton College in the fourteenth century.[3.33] They also made contributions to trigonometry and the construction of instruments. The most important of these was that of Levi ben Gerson of Provence (1288–1344), who popularized the cross staff, a kind of primitive sextant, which served the navigators of the voyages of discovery of the fifteenth and sixteenth centuries. It is interesting that apparently the first serious scientific work in English is the recently discovered *Equatorial Planetarie*, [3.52] a mechanical device for predicting planetary positions, described though not invented by Geoffrey Chaucer (*c.* 1340–1400), whose *Treatise on the Astrolabe* 'for little Lewis my son' has long been known.[3.23] There was no radical revision of astronomy, for though the opposition – impetus – school of Albert of Saxony (*fl. c.* 1357), Oresme and, most clearly, Nicholas of Cusa dared to suggest that it was the earth and not the heavens that turned daily, they did so on philosophic grounds. They were not astronomers themselves and professional astronomers continued to follow Ptolemy till well into the seventeenth century.

THE LIMITATIONS OF MEDIEVAL SCIENCE

Though the contribution of medieval Christendom to science may have been unfairly ignored in the past, the danger today is rather to exaggerate its importance to the extent of making the whole history of science unintelligible. The significant fact is that as a live tradition it flourished only in the twelfth and thirteenth centuries and had by the early fifteenth lapsed into obscure pedantry which justifies and explains the contempt of the men of the Renaissance for Gothic barbarism.[3.3] This fact, coupled with the practical identity of the subjects treated and the methods used by the schoolmen with those of Islamic science, points to the conclusion that medieval science as a whole must be treated as the end rather than the beginning of an intellectual movement. It was the final phase of a Byzantine–Syriac–Islamic adaptation of Hellenistic science to the conditions of a feudal society. It arose as a consequence of the breakdown of the old classical economy and was in turn to decay and vanish with that of the feudal economy that succeeded it.*

It is unfair to expect more of such a science than what was demanded from it in its time. Both for the Muslim and the Christian natural science had a share, and not a very important one, in the great task of justifying the divine order of the universe, whose main features were given by revelation and supported by reason, that is by abstract logic and philosophy. Robert Grosseteste, probably the medieval scholar with the finest mind and with the greatest influence on the development of medieval science, thought of that science essentially as a means of illustrating theological truths. His study of light and his verification by actual experience of the refraction of lenses were undertaken because he conceived of light as analogous to the divine illumination.[3.26]

Those who thought otherwise in the Middle Ages, and there were very few of them, were likely to be prosecuted for heresy or at best ignored. Here again Grosseteste's pupil, Roger Bacon, the most authentic voice from that time preaching a science for the service of man and prophesying the conquest of Nature through knowledge, proves how far we have come from the medieval outlook. Though he predicted motor ships, cars, and aeroplanes, and an alchemical science 'which teaches how to discover such things as are capable of prolonging human life', his interest in science was essentially theological. For him scientific knowledge is only part, with revelation, of an integral wisdom to be contemplated, experienced, and used in the service of God.

The overriding need was to justify the truths of Christianity, as pointing to the true *end* of human existence on earth. No mundane knowledge could be compared to that of the scheme of salvation to which the Church, with its sacraments and traditions, held the key. It was such considerations that directed medieval thought to the ordering of all knowledge and experience to build one majestic world-picture containing in essence all that it was important for man to know. This encyclopedic tendency reached its height in the Middle Ages, not only in the complete logical scheme of Thomas Aquinas' *Summa*, but also in other works containing more general information like those of Bartholomew the Englishman (*fl. c.* 1230–40) and Vincent of Beauvais (d. *c.* 1260) whose *Speculum Majus* was not equalled in length until the French *Encyclopédie* of the eighteenth century (p. 526).

THE MEDIEVAL WORLD-PICTURE

It is necessary to say something here of this medieval world-picture if only because modern science arose largely out of the attempt to supersede it, and still bears many of the signs of the struggle. The main characteristics of the Graeco-Arabic–medieval system were those of completeness and of hierarchy. The ethereal, cosmological scheme of Aristotle (p. 200) and the Alexandrian astronomers (p. 216) had become a rigid, theological–physical world, a world of spheres or orbs – the spheres of the moon and the sun; the spheres of the planets; above all the great sphere of the fixed stars beyond which lay heaven; and, as a theologically necessary counterweight, the underworld, the circles and pits of hell so grimly described in Dante's *Inferno*. The world was ordained as one of rank and place. It was a compromise between the Aristotelian picture of a permanent world and the Jewish and Christian picture of a world created by one act, only to be destroyed by another. It was an interim world which, though it had its own rules, was there merely as a stage for the playing out of each man's life on which depended his ultimate salvation or damnation.

86. The hierarchy of the Church, with God at the head, the Pope below with angels, kings, priests, laity and devils. From Elias Ashmole, *Theatrum Chemicum Britannicum*, London, 1653.

HIERARCHY

The hierarchy of society was reproduced in the hierarchy of the universe itself; just as there was the pope, bishops, and archbishops, the emperor, kings, and nobles, so there was a celestial hierarchy of the nine choirs of angels: seraphim, cherubim, thrones; dominations, virtues, and powers; principalities, archangels, and angels (all fruits of the imagination of the pseudo Dionysius). Each of these had a definite function to perform in the running of the universe, and they were attached in due rank to the planetary spheres to keep them in appropriate motion. The lowest order of mere angels that belonged to the sphere of the moon had naturally most to do with the order of human beings just below them. In general there was a cosmic order, a social order, an order inside the human body, all representing states to which Nature tended to return when it was disturbed. There was a place for everything and everything knew its place. The elements were in order – earth underneath, water above it, air above that, and fire, the noblest element, at the top. The noble organs of the body – the heart and lungs – were carefully separated by the diaphragm from the inferior organs of the belly. The animals and the plants had their appropriate parts to play in this general order, not only in providing man with necessities, but even more by furnishing him with moral examples – the industriousness of the ant, the courage of the lion, the self-sacrifice of the pelican. This tremendous, complex, though ordered cosmos was also ideally rational. It combined the most logically established conclusions of the Ancients with the unquestionable truths of Scripture and Church tradition. The schools might differ on some of its details but none doubted that it was substantially a true picture. The essential problem had, as it seemed, been solved for all time. It was possible to have a universe which was at the same time practical, theologically sound, and eminently reasonable.*

6.6 The Transformation of Medieval Economy by New Techniques

In the light of this it is easy to see how an attack on any part of the universal picture was considered to be something much more serious than a mere intellectual adjustment, but rather an attack on the whole order of society, of religion, and of the universe itself. It was therefore necessary to resist it with all the power of Church and State. The medieval system of thought was necessarily conservative and if it had been left to itself it would probably have been conserved to this day. But it was not left to itself. However much the medieval system of thought might tend to be static the medieval economy could not stay still.[3.12]

The feudal system, as has already been explained (pp. 287 f.), contained the seeds of its own transformation. Greater trade and improved techniques of transport and manufacture drove relentlessly towards a commodity and money economy in place of one based on prescribed service. It was the technical aspect of this economic revolution that was to be the decisive factor in creating a new, progressive, experimental science to take the place of the static, rational science of the Middle Ages. It was to present the men of the Renaissance with situations and problems that the old knowledge was inadequate to deal with.

These intellectual adjustments consequently belong to the later period, but the essential technical changes themselves took place during the Middle Ages and indeed represent their most significant contribution to the scientific civilization of the future. In such an apparently well-ordered and static society these technical changes remained for long unrecognized because they were for the most part beneath the notice of the clerical chroniclers, though they appear prominently enough in manorial accounts and lawsuits. We have one precious document in the notebook of a master mason, Villard de Honnecourt (c. 1250),[3.5] containing accounts and drawings of many mechanical devices. Very

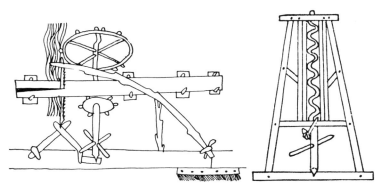

87 a, b. Two mechanical devices designed by Villard de Honnecourt: a water-driven saw and a screw-jack. Redrawn from a page of his 'Album'. From *The Renaissance Engineers* by Bertrand Gille, Lund Humphreys, London, 1966.

few of the medieval scholars mentioned technical matters and fewer still tried to understand them. How exceptional such an interest was is shown by Roger Bacon's eulogy of Peter the Pilgrim:[3.26]

He knows natural science by experiment, and medicaments and alchemy and all things in the heavens or beneath them, and he would be ashamed if any layman, or old woman or rustic, or soldier should know anything about the soil that he was ignorant of. Whence he is conversant with the casting of metals and the working of gold, silver, and other metals and all minerals; he knows all about soldiering and arms and hunting; he has examined agriculture and land surveying and farming; he has further considered old wives' magic and fortune-telling and the charms of them and of all magicians, and the tricks and illusions of jugglers. But as honour and reward would hinder him from the greatness of his experimental work he scorns them.

Such an ideal was very far, however, from the aspiration of the school-men, who paid scant attention to matters with so little bearing on salvation or preferment. The Renaissance humanists, who thought all good things came straight from Greece or Rome, for their part deliberately ignored them. They were in revolt against the whole achievement of the Middle Ages, which they stigmatized as barbarous and Gothic.

MEDIEVAL ARCHITECTURE

Yet we, who are no longer fighting a life-and-death struggle against feudalism, have only to look at the development of that Gothic architecture, from the dark massiveness of the Norman to the luminous lightness of the perpendicular, to see that those three centuries span a world in rapid technical advance. Architecture was indeed the greatest and most characteristic expression of medieval technique and thought. It was, however, a purely technical rather than a scientific achievement. The marvellous construction of vault and buttress, far more daring than anything the Romans or Greeks attempted, were the result of a series of *ad hoc* solutions to practical difficulties. Theory did not enter into them at all; nor could it, for the theory of the arch, apart from the working knowledge of it, was only discovered in our time. For the same reason medieval architecture contributed little, directly or indirectly, to the advance of science.* It was different with other innovations, some of which, like the compass or gunpowder, were to furnish the bases of the new science, while others, like horse harness and the sternpost rudder, were to affect science indirectly through the improvement in productivity they brought about. [1.28; 1.29; 2.51; 2.52]

88. A rubbing from the tomb of Wu Liang, 147 A.D., showing improved horse harness and shafts that made a great difference to the load that a horse could draw. (Note also harness shown in plate 79.)

TECHNICAL INNOVATIONS FROM THE EAST AND CHINA

The technical advances of the Middle Ages were made possible by the exploitation and development of inventions and discoveries which, taken together, were to give Europeans greater powers of controlling and ultimately of understanding the world than they could get from the classical heritage. Significantly, the major inventions – those of the horse-collar, the clock, the compass, the sternpost rudder, gunpowder, paper, and printing – were not themselves developed in feudal Europe. All seem to have come from the East and most of them ultimately from China.

As we come to know more about the history of science in China (and there Dr Joseph Needham's great study of the origins and history of Chinese techniques and science will be invaluable)[3.8], we are beginning to see the enormous importance for the whole world of Chinese technical developments. Already enough is known to show that the whole concept of the superiority of Western Christian civilization is one based on an arrogant ignorance of the rest of the world. Transmission is always difficult to prove, but the fact remains that many inventions only appearing in the tenth century or later in western Europe were fully described in China in the very first centuries of our era.

What still requires to be explained is why this early technical advance in China, and to a lesser extent in India and Islamic countries, after a promising start came to a dead stop before the fifteenth century, and why it resulted in the formation of Oriental civilizations with a high but static technical level. The reason given by Dr Needham as especially applicable to China is the rise of a bureaucracy – the Mandarins – with a literary education, having no interest in improving technique and being very concerned with keeping down the merchants, who alone could have driven techniques forward by opening up new markets.*

It was precisely this that was to happen in Europe. The new inventions, in the measure in which they came to be used, set in motion a revolution in technique which contributed in a cumulative way to the breakdown of feudal organization through increased productivity and trade. Better means of agricultural production in the villages meant more surplus to exchange. Better transport of bulk goods relieved the need to produce everything from land more suited to a particular crop, and thus indirectly increased productivity. For instance, whole districts round Bordeaux were given up to wine-growing in the thirteenth century, for wine was the first bulk cargo, as witness our present heavy unit of weight, the ton – originally the weight of a tun or a barrel of wine. Trade in turn enhanced the importance of merchants and thus of the towns, and handicraft industry began to grow in town and country.

The characteristic of medieval economy most significant for the future was that the towns did not dominate the country. The feudal system maintained this independence, and the absence of slaves prevented the rise of factories on the classical pattern. The industry which arose from the new inventions was spread over hundreds of villages. This was particularly so when mills became a main source of power, for they strung out production along fast streams and windy ridges. Mining and smelting had necessarily to be scattered, country industries. This rural location increased the chronic labour shortage already mentioned, and put a premium on mechanical ingenuity. Moreover, by going to the country the restrictions imposed by town guildsmen on new processes which would put them out of work could be evaded.

THE NEW HORSE HARNESS

Of the inventions listed, the horse-collar and the mill were essentially more efficient ways of transmitting power. Of these the first had the most immediate effect; by substituting a collar, pulling on the shoulders of the horse, for a band across his breast, which constricted his windpipe, the permissible tractive effort was increased fivefold.[2.52] This innovation,

89. An animal-powered grinding mill from the 1620 German edition of *Le Diverse et Artificiose Machine* by Agostino Ramelli (*c.* 1530–90). (See plate 81.)

coming from seventh-century China, reached Europe early in the eleventh century. Its immediate results were that horses could take the place of oxen at the plough, and in addition acres of land unsuited to ox-ploughing could be cultivated. At the same time the horse-cart took the place of the ox-cart. The simultaneous introduction of nailed horseshoes put the horse on the road for pack and wagon transport. The advantages of the new horse harness accrued in the first place to the countries of the Franks and Normans and began to make the area round the North Sea and the Channel, already favoured by good soil and a drought-free climate, a major centre of production. The surplus of corn, fish, hides, raw wool, and cloth – the main commodities of the new heavy merchandise – could then be exchanged at great fairs, such as those of Champagne, for the more finished but lighter products of the East and the South.

THE WATER-MILL AND WINDMILL

The actual invention of water-mills belongs to the classical period; one is described by Vitruvius (*c.* 50 B.C.). The mill, however, has a right to be

90. The trip hammer, a device for turning rotary into reciprocal motion. From Fludd's *Utriusque Cosmi . . . Historia*, Oppenheim, 1617–19. (See plate 60.)

considered as a medieval device because it was only in the Middle Ages that it came to be widely used. Roman mills were few; streams were not very suitable for them, and Mediterranean slaves could always be found to do the work. In contrast, the mill was, from the start, an integral feature of feudal economy. A mill and a miller were to be found in almost every manor (5,000 of them are listed in the Domesday Book) and the lord made full use of his right to demand that all his serfs have their corn ground at his mill.

Nor were mills limited to grinding corn; they opened the way to a more general use of power.[3.24] Wherever steady or repeated applications of force were necessary to which work could be brought – for the mill was intrinsically static – mill mechanism could be adapted. For the conversion of rotary into reciprocal motion came two devices, both apparently from China, the trip-hammer and the crank;[3.8] the latter is important because, unlike the trip-hammer, it can also be used to convert reciprocal into rotary motion. Windmills, apparently from Persia, reached Europe about 1150. Mills were used for fulling cloth, blowing bellows, forging iron, or sawing wood, but not until the Industrial Revolution (p. 521) for the equally arduous but more scattered tasks of spinning, weaving, or threshing. The very fact of the use and rapid development in Europe of mills for so many purposes bears witness to the shortage of labour and to the connexion between this and technical and scientific development.

Wind and water-mills needed to be made and serviced, a task beyond the skill of most village smiths. So there grew up a trade of *millwrights* who went about the country making and mending mills. These men were the first mechanics in the modern sense of the word. They understood how gears could be made and how they worked as well as the management of dams and sluices, which made them hydraulic as well as mechanical engineers. They were the repositories of ingenuity from which the Renaissance, and even more the Industrial Revolution which followed it (p. 547), drew the craftsmen who alone could have put into practice the ideas of the new philosophy.

THE CLOCK AND THE WATCH

The mechanics also had a hand in the development in medieval Europe of the present form of mechanical clock. The clock, as its name implies, was originally just the bell (cloche) rung to mark the hours of service – later all the hours. It was rung by a watchman using an hour glass. Somewhere in the eleventh century an ingenious mechanism, the verge and folliot, which imparted a to-and-fro motion to the clapper, was devised. All the watch had to do was to release a weight which, through

91. The early fourteenth-century weight driven clocks of western Europe were crude by comparison with those that were to follow. They were the product of the blacksmith rather than the instrument maker, yet this clock mechanism from Wells Cathedral (now separated from the face and the jacks, which are still in the Cathedral) is a fine example of early workmanship. The mechanism, now in the Science Museum, London, contains some additions of a later period.

a train of *clockwork* (essentially a lighter form of millwork), struck the appropriate hour. It occurred to some millwright or monk that the same mechanism working over and over again could be used to tell the time itself, thus making a mechanical *watch* – as it is still known in the trade – and eliminating the watchman. So the mechanical *clock*, which included the watch, was born, the prototype of modern automatic machinery – self-regulating as well as self-moving.

Timepieces are of course of great antiquity. The Arabs improved greatly on the Greek water-clocks and made them the basis of many complicated and automatic devices; but these were operated by floats and cords and lacked the precision and the force of trains of gear wheels. We now know, however, that cog-wheel gearing has a far greater antiquity both in Greece and China.* The clock can no longer be claimed as a European invention, though it was most developed there. Clocks were objects of prestige, rather than of use. They were the pride of towns or cathedrals, but the rare trade of clockmaker and afterwards of watchmaker was in the Renaissance to become for science what the millwright was to be for industry – a fruitful source of ingenuity and workmanship.[3.42]

THE MARINER'S COMPASS

The observation of the directive power of the earth's magnetism on a natural magnet or lodestone must have been one of the most difficult, as well as the most important, of scientific discoveries. There seems little doubt that the directive property of a pivoted lodestone was known to the Chinese several centuries before we have any record of its use elsewhere.

The discovery seems to have been made, according to Dr Needham, [3.40] as a by-product of geomantic divination – a practice of throwing objects on a board and foretelling the future from the way they lay. These practices still continue and have, incidentally, given us most table games, including dice, cards, and chess. One object was the sign of the North, the Great Bear or Dipper, represented in the form of a spoon. Such spoons cut from lodestone – one of the five sacred stones – would always point in one direction. It was discovered before the sixth century that this pointing property was also possessed by pieces of iron touched by the lodestone or even allowed to cool when pointing north and south. A water compass in which such a piece of iron was supported on wood is fully described in the eleventh century, but was probably known long before. This is the traditional Chinese compass, its association with the divining boards being shown by the symbols on its frame. How it passed to the West is still a mystery. There is a reference to it as already well known in a twelfth-century saga. The pivoted needle and the card with the windrose seem to be Italian inventions of the thirteenth century.[3.11]

The slow development of the compass after its first discovery bears all the marks of traditional, technical improvement; but science was early invoked to explain its action. The first original scientific work of western Christendom was *Epistola de Magnete* (1269), the work of Peter the

Pilgrim (de Mericourt), the contemporary of Roger Bacon, who admired him as the greatest and most practical scientist of the age (p. 310). It shows a great independence of thought and a capacity for planning and carrying out a sequence of experiments. From this work – after a long interval – were to stem the researches of Norman and Gilbert (pp. 434 ff.), from which was to come the whole of the theory and practice of magnetism and electricity. Not only that, but the influence of the magnet on the compass was to provide the real scientific basis of the doctrines of influence and inductions which had previously been purely magical. Even more important, it was to furnish a working model of the doctrine of attractions which permeated the whole of science, and which was to be the guiding star of the great synthesis of Newton.

THE STERNPOST RUDDER

The sternpost rudder also apparently came from China. The Chinese junk is radically different from the ship in that, while the latter was developed from the original dug-out canoe by building up the sides around a central keel, the former is derived from a bamboo raft by lifting up bow and stern.[2.51; 3.31] It has no keel and the natural place for the rudder is the middle of the stern. In Europe the central rudder was more difficult to attach because of the old sloping form of the keel at the

92. Warships firing broadsides. The ship on the left has a sternpost rudder. From a sixteenth-century woodcut.

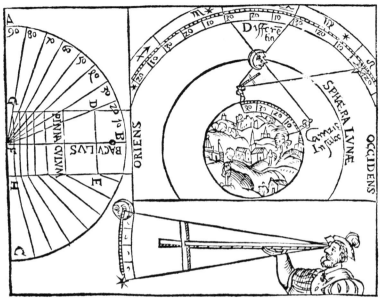

93. The cross-staff was much used in medieval and Renaissance times. It came to western Europe via the translations of Arabic texts in the fourteenth century by the Jewish scholar Levi ben Gerson, and was often known as 'Jacob's staff'. It was improved and considerably modified in the seventeenth century. Both astronomical and navigational uses are shown here. From Apian's *Cosmographia* (see plate 45).

stern, and a steering oar fixed to the *star*board was used, but once this was done by adding the vertical *stern*post somewhere in the thirteenth century, it made the deeper-keeled European vessels, based on Viking models, much better sailers. A course could now be held with sails set closer to the wind. This in turn led to the development of the fore-and-aft sail from the older lateen sail. Winds astern had no longer to be waited for and voyages could be made in rougher weather.

The two navigational inventions, the compass and the sternpost rudder, were to have an effect at sea of importance comparable to that of the horse harness on land. Their use made open sea voyages feasible, and such voyages largely took the place of the roundabout coasting of earlier times. They threw the oceans, for the first time, open to exploration, war, and trade, with enormous and rapid economic and political results.

NAVIGATION

The scientific consequences of the development of navigation were to be of critical importance. Open-sea navigation, even in the Mediterranean, required astronomic observations and charts, and gave a direct stimulus to the development of an astronomy capable of accurate predictions, of a new quantitative geography, and of instruments suitable for use on shipboard. Ocean navigation further raised the urgent problem of finding the longitude, at which all the great astronomers of the seventeenth century were to try their hand. The need for compasses and other navigating instruments brought into being a new skilled industry, that of the card and dials makers, whose subsequent influence on science, particularly in setting higher and higher standards for accurate measurement, was enormous. Many scientists, including Newton himself, were instrument makers, and one instrument maker, Watt, was to have a revolutionary effect on industry and on science.

LENSES AND SPECTACLES

The discovery of lenses already described (p. 278), led by 1350 to the invention of spectacles, apparently in Italy. Their use gave still further impetus to the study of optics. Grosseteste, Roger Bacon, and Dietrich of Freiburg explained the action of a lens both in focusing light-rays and magnification.[3.26] The demand for spectacles gave rise to the trades of the lens grinders and spectacle-makers. They were able to flourish only owing to the availability of cheap clear glass (*n.* p. 310). To one of them, traditionally Lippershey in 1608, we owe the invention of the telescope, and it would seem, at least at that stage, that the casual combination of lenses, possible only in a spectacle-maker's shop, was more fruitful than any theoretical conjectures on the magnification of images.

GUNPOWDER AND CANNON

Of all the inventions introduced to the West in the Middle Ages, it was the most destructive – gunpowder – that was to have the greatest effect politically, economically, and scientifically. The original invention has been claimed for the Arabs and the Byzantine Greeks, but the balance of evidence is for a Chinese origin. The key to its operation is the addition of a nitrate (nitre) to make combustible substances burn without air. Nitre occurs naturally in some salt-pans and also in overmanured ground. Either it was first used by chance, in firework compositions, or possibly it was noticed that using it instead of soda (natron) as a flux with charcoal led to a bright flash and mild explosion. In China for some centuries it was used merely for fireworks and rockets.

The military importance of gunpowder started when it was used in the cannon, perhaps derived from the fire-tube of the Byzantines, but more probably from the bamboo cracker of the Chinese. The very name of the barrel of a cannon indicates its primitive construction from iron staves hooped together. The cannon, and the hand-guns which soon followed them, were effective in war not so much because their range or power exceeded that of the old catapults and ballistae, but because, for all their clumsiness and cost, they were far cheaper and more mobile. Their use in battle and sieges initiated a technical revolution in warfare, comparable only with that which took place at the beginning of the Iron Age 3,000 years before.[2.65]

Against foes without it, gunpowder, with cannon and muskets, gave practical invincibility and thus put 'civilized' man in a position of effective superiority against far more numerous 'natives'. But even among the civilized it enormously altered the balance of power. Once cannon came

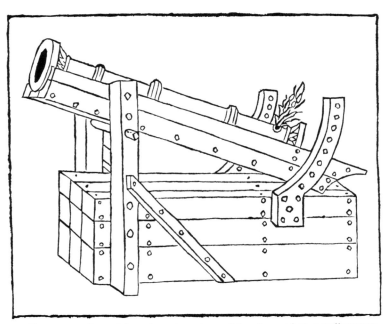

94. The cannon, formed from iron staves hooped together and on a rudimentary mounting. From *De Re Militari* by the military engineer Roberto Valturio (*c.* 1450–*c.* 1480), published in 1483.

in they became a necessity for victory, and from an economy turned into a new expense of war. Only wealthy republics or kings backed by merchants could command sources of metal and the technical skill to fashion it into cannon. This fact broke the independence of the land-based aristocracy as surely as their castles were battered down by cannon balls. The triumph of gunpowder was the triumph of the national State and the beginning of the end of feudal order.

At sea the effect of gunpowder was no less important. Used in naval guns, mounted in ships directed by the new astronomy and the compass, gunpowder was to make the western Europeans supreme over the sea-ways of the world from that time to the middle of the present century. It enabled Europeans to stamp their pattern of culture on others, originally by no means inferior, culturally or militarily. More immediately it enabled them to concentrate the accessible wealth of the world in their hands, and so to possess the accumulation of capital which financed the Industrial Revolution.

THE SCIENTIFIC CONSEQUENCES OF GUNPOWDER –
CHEMICAL AND PHYSICAL

Ultimately, however, it was the effects of gunpowder on science rather than on warfare that were to have the greatest influence in bringing about the Machine Age. Gunpowder and the cannon not only blew up the medieval world economically and politically; they were major forces in destroying its systems of ideas. As Mayow put it, 'Nitre has made as much noise in philosophy as it has in war.' In the first place they were something new in the world – the Greeks did not have a word for them. In the second place the making of gunpowder, its explosion, the expulsion of the ball from the cannon, and its subsequent flight furnished problems the practical solution of which led to a search for causes of a new kind and the creation of new sciences.

Whatever the origin of gunpowder, the essential ingredient – nitre (potassium nitrate) – could have been produced only as the result of a careful study of the separation and purification of salts, probably in connexion with alchemy. Wherever it had to be made it turned attention to the phenomena of solution and crystallization. Moreover, to explain the explosion of gunpowder taxed medieval chemistry and physics to the utmost. It was clearly an action of fire, but unlike all other terrestrial fires it did not require air. This led to the speculation that the air was provided by the nitre and conversely that air contained nitre, or at least a nitrous spirit (*anima*). It thus became the model for all subsequent attempts to explain combustion and with it breathing, that animal necessity for air. Ultimately, after four centuries of argument and experiment, it was to lead to the discovery of oxygen and with it to the whole of modern chemistry (pp. 618 f.).

The force of the explosion itself, and the expulsion of the ball from the barrel of the cannon, was a powerful indication of the possibility of making practical use of natural forces, particularly of fire, and was the inspiration behind the development of the steam-engine (pp. 576 ff.). Later we shall see how the machinery developed for the boring of cannon (p. 582) was to be used in making accurate cylinders which gave the early steam-engines a chance to prove their efficiency.

Finally, the movement of the cannon ball in the air – ballistics – was to be the inspiration for the new study of dynamics. The classical scientists had studied bodies at rest, or bodies acting on each other with relatively steady forces. The new world was to consider the problem of bodies in violent motion, and on this basis was to found a new and much more comprehensive mechanics.

Impetus theory came long before the cannon, but the interest in the flight of shot focused a new attention on it. The new mechanics differed from the classical in one vitally important respect: it depended on, and in turn generated, mathematics – it was quantitative and numerical.

DISTILLATION AND ALCOHOL

The first preparation of strong spirits of wine was made in Europe in the twelfth century, although most of the steps leading up to it had already been taken in their development of distillation by the Arabs. The last decisive step was probably made in Salerno, whose medical school was already famous. It had been founded in the ninth century, and eventually absorbed the best of Arab science from that melting-pot of Greek, Arab, and Norman culture – Sicily. As the distillation of perfumes and oils was already known, alcohol was probably hit on by accident in the course of some medicinal preparation. The clue to its preparation was to cool the still-head or alembic sufficiently to condense the alcohol as well as the water.[2.28] The resulting distillate was first drunk as a rare medicine and its cordial properties were noted. Soon it could be made strong enough to burn, which added much to its prestige. In the fourteenth century, Raymond Lull is alleged to have distilled wine with quicklime and produced nearly absolute alcohol. The name is a misnomer; the Arabic term really applies first to eye-paint and then to any fine powder. The great demand for alcohol – fire water, usquebaugh, whisky, burnt wine, brandy wine – came only with the Black Death in the fourteenth century. It was believed that those who drank it regularly would never

95. Large brick still from a woodcut in one of the first books to deal with the severely practical aspects of alchemy – applied chemistry, in fact – the *Buch zu Distillieren* by Hieronymus Braunschweig, published in 1512 at Strasbourg.

die, hence the name *aqua vitae*. After that it got right out of the control of the doctors and began to be produced in quantity, as attested by the numerous laws promulgated against its use. Alcohol gave rise to the first scientific industry, that of the distillers, the foundation of the modern chemical industry.

The social and scientific results of the preparation of alcohol were manifold. The most obvious, the effects of drinking it and the craving it produces, were of no great social importance in Europe, but in heathen parts alcohol was second only to gunpowder in its civilizing mission. (Manhattan Island was purchased by the Dutch from the Indians in 1626 for three barrels of rum. The name means 'the place where we got drunk'.) For science, alcohol had a double significance – chemical and physical. The capture of the spirits of alcohol gave great impetus to applying the same method to other substances. Now, the far more efficient, water-cooled condensers that were produced by the industry meant that other volatiles, such as ether, might be condensed. The still and condenser supplemented the alembic and retort as the chief laboratory apparatus and made organic chemistry possible.

The physical processes of distillation, particularly the strange transfer of heat from the fire to the condenser water, proved very difficult to understand. It was left for Black (pp. 580 f.) in the eighteenth century to draw from it the doctrine of latent heat – the beginning of thermodynamics. In turn it was from this doctrine that Black's instrument maker, Watt, invented the separate condenser and produced the first thermally efficient engine.

PAPER

The last two technical introductions from the East, which were fated to have a far greater effect in the West than in their land of origin, were the linked inventions of paper and printing. The need for a writing material cheaper than the very expensive parchment, became more and more urgent with the spread of literacy. The process of paper-making was developed originally in China, based on vegetable fibres. It was already being used there as a cheap writing material in the first century B.C. It was introduced to Europe via the Arabs in the twelfth century. In Europe linen rags provided the basis for the first paper of quality, since then unexcelled. Paper turned out to be so good and cheap that its increased availability led in turn to a shortage of copyists and hence to the success of the new method of copying which printing provided.

PRINTING

The technique of printing is not a very difficult one to invent or to practise. In fact in seals, rubbings, and stampings it has been used since the very earliest times. Its rapid spread in Europe was an example of a social and organizational need making use of and further developing a technical device. Before a need can be effective it must be felt to exist. But the

96. Paper, an innovation from the East, coupled with the invention of printing, transformed the dissemination of knowledge. A paper mill, using water power in the form of an undershot waterwheel. From Georg Andreas Böckler, *Theatrum Machinarum Novum*, Nuremberg, 1673.

particular need that brings the technique into existence is not necessarily the main one that the new technique ultimately comes to serve.

Even in the late Middle Ages few people were aware of the need for a large quantity of paper books. In fact, printing would probably not have been developed in the first place merely for literary purposes. The full value of printing is felt only when large numbers of cheap copies of one text are needed. Consequently it is not surprising that it first arose in the East for the reproduction of Taoist or Buddhist prayers, where quantity is a definite spiritual advantage, and later for the printing of paper money, which also implies large numbers. In the West, oddly enough, it was another use, the development of playing-cards, originally a form of divinatory magic, that gave rise to the need for large-scale block printing, with papal indulgences, prayers, and sacred images not far behind.

CHEAP BOOKS, RELIGION, AND THE NEW LEARNING

Printing with movable wooden type was originally a Chinese invention of the eleventh century. Movable metal types were first used by the Koreans in the fourteenth century. It was introduced into Europe in the mid fifteenth century and spread extraordinarily rapidly, first for prayers and then for books. The new, cheap, printed books promoted reading and thus created the need for more books, so setting off a kind of explosive or chain reaction. Naturally the printers first concentrated on producing larger numbers of the books that had been most in demand as manuscripts. The original centre of interest was in religion and particularly in the Bible, whose printing and dissemination to the rising middle classes fell in with the new trend of emancipation of thought from Church control that was to lead to the Reformation. A close second was literature and poetry, both ancient and modern, for the delight of the now cultured aristocracy and upper bourgeoisie of the Renaissance.

Later still, largely in the sixteenth century, printing was to be the medium for great technical and scientific changes by its setting out at large, for all to read and see, descriptions of the world of Nature, particularly of its newly discovered regions, and also, for the first time, of the processes of the arts and trades. Hitherto the techniques of the craftsmen had been traditional and never written down. They were passed on from master to apprentice by direct experience. Printed books made it first possible and then necessary for craftsmen to be literate. Their descriptions of technical processes, and even more their illustrations, helped to bring about for the first time close relations between the trades, the arts, and the learned professions (pp. 388 f.).

6.7 The Development of Late Medieval Economy

This discussion of the importance of printing has brought us beyond the limits of the Middle Ages but, before passing to a consideration of the revolution in science of the Renaissance, it is necessary to assess the effect of these and other technical advances taken together on the economy and ideas of the late Middle Ages. Over the countryside as a whole the combined effects of improved production and transport were to increase the gross surplus of the village and consequently the amount of manufactures that could be consumed there.

All over Europe, though the dominance of feudal lords was not yet shaken, wealthy peasants and urban workers strengthened their position and began to provide a large-scale market. This in turn stimulated the manufacture of goods, particularly of semi-luxuries like wine and good cloth (rough cloth was still spun and woven at home), and the production of extra food, like salt fish, and also of metals, particularly iron for tools and weapons. These manufactures, though carried on more often in the

97. The Steel Yard, in Thames Street, was the London Headquarters of the Hanse merchants or Esterlings. This engraving 'from an Antient view' was executed by Bartholomew Howlett (1767–1827).

country as a part-time peasant occupation, were dominated by town merchants. By the mid thirteenth century, which may be thought of as the turning point of the Middle Ages, the rich town merchants had acquired through their dominance of the guilds a monopoly position which they used in order to buy cheap and sell dear.

These town oligarchies were often in violent opposition to each other, sometimes to the extent of war. Towards the latter part of the Middle Ages they began to appreciate the value of co-operation for the common exploitation of less-developed territories. The most famous of these associations was the North German Hanse, centred on the exploitation of the Baltic trade. From about 1358 to 1550 it virtually ruled the old Viking strongholds of Scandinavia. The Hanse had its own navy and maintained factories in other towns, from the Steelyard in London to Novgorod, with extra-territorial rights. By concentrating on buying up raw materials in outlying countries and selling them as finished goods, it depressed the development of industry outside its own cities.

This extension of the range of action of city leagues postponed but did not remove the causes of conflict inside the cities. Nor was it possible for foreign merchants to maintain commercial domination indefinitely in the face of the growth of native resources. Britain, for instance, was, up to the fifteenth century, a country exporting raw wool which was worked up in Flanders and Italy.[3.50] Financially it was dominated by Lombards, Florentines, and Hansards. It had become, in fact, a semi-colonial country though, like the North American colonies in the eighteenth century, one with such resources that its economic independence was only to be a matter of time. Indeed its emancipation started with the development of the domestic weaving of wool as early as the fourteenth century.

In the most advanced medieval cities, those of Italy and the Low Countries, the rule of the wealthy guildsmen provoked revolts of the craftsmen, like that of the Ciompi in Florence in 1378 and of the weavers of Bruges, Liège, and Ghent from 1302 to 1382. Though these revolts succeeded, they did not lead to the achievement of city democracies of the Greek type, because the medieval cities were set in a far more developed and populous feudal countryside. Instead, the final result of struggles inside or between cities was to strengthen either the feudal kings or the merchant princes and hired captains (*condottieri*), who seized power in Italy. This was to lead to the establishment of the nation States of the Renaissance, still feudal in essence but centred on the towns. It was only in a later period that the capitalist system was to grow from this bourgeois nucleus.

98. The medieval guild was a closed community, bound together by a common aim. In London it carried on in a modified form in the City Livery Companies. The front of the Mercers' Hall, Cheapside, from a drawing and engraving by T. Busby.

COMMERCE AND MATHEMATICS

It is to the cities, therefore, that we must look for the development of ideas and particularly of science in the later Middle Ages. Here were growing up a new lay intelligentsia, good Christians but largely independent of, and in some degree in opposition to, the Church, which was still by far the greatest landowner, and firmly attached to the feudal system. At first, however, their interests hardly clashed, for the new *bourgeoisie* were interested more in profit and display than belief. Commercial arithmetic, fine craftsmanship, and art concerned them

99. 'The Suitors surprising Penelope', a fresco by Bernadino Pinturicchio (1454–1513). Penelope is working at a contemporary two-heddle loom, the maidservant (left) winds a shuttle-spool from a ball of wool, and the ship in the background is a carrack. The artist has given much natural and technical detail, and the subject of his fresco – 'Scenes from the Odyssey' – was secular and in contrast with the religious subjects previously adopted.

far more than the disputes of the schools. It was only later, when they found the Church an obstacle to their increasing wealth and power that they were to become the most ardent advocates of reform.

The Arabic numbers introduced by Leonardo Fibonacci in 1202 found their main use in commercial accountancy. Within a few decades the four rules of arithmetic, hitherto a mystery confined to a handful of mathematicians, became a necessary training for every merchant apprentice, incidentally creating a large body of persons able to appreciate mathematics. The result was symbolic algebra and the signs of $+$ and $-$, originally checkers' marks for over and below weight. It was the same mercantile interest that first kept up and later improved astronomical tables and new maps for the benefit of navigation.

ART AND SCIENCE

The increased wealth of the merchants gave a new impulse to art and at the same time changed its objects and its style. Though still expressed in a religious form, it was no longer the Church art of the early medieval period as embodied in the Gothic cathedrals. Illustrations from Nature took the place of theological symbolism. Art was becoming at the same time more secular and more naturalistic. Much of the surplus accumulated by the merchants was spent on mansions and pictures, partly for pleasure, partly for prestige.[1.62] The number of craftsmen multiplied, and their techniques were continuously improved. In textiles, pottery, glass, and metal-work there was ample incentive and opportunity for practical research in the properties of matter, physical and chemical. This was to provide the material basis for the revival of science. The stage was set for the full flourishing of the Renaissance.

6.8 The Achievement of the Middle Ages

The legacy of the Middle Ages was essentially economic, technical, and political. Its intellectual contribution was not so lasting. For whereas the foundation laid by feudal economy modified by urban trade was able to support the further advances of the Renaissance and the Industrial Revolution without any breakdown, the ideas of the Middle Ages had to be ruthlessly scrapped before a new scientific philosophy could take their place. This is not to disparage the enormous intellectual effort of the scholars of the Middle Ages involved in recovering and

100. Arithmetic as depicted in the *Margarita Philosophica* by Gregory Reisch (see plate 5). An abacus-style counting table is shown, also 'arabic' numbers and fractions. The allegorical figure of arithmetic holds out the wisdom of the ancients.

absorbing the elements of classical science. However, for the reasons already discussed, they were as incapable as the Arabs before them of advancing beyond the limits that had been reached by Aristotle nearly 2,000 years earlier.

The medieval contributions were certainly more finished than those of the Arabs. They had established the principles of the scientific method. Robert Grosseteste at the outset of the period had stated the double method of *resolution* and *composition* or of *induction* and *deduction* as clearly as Newton was to put it 500 years later.[3.26] But method without either the desire or the means to use it is almost worse than useless. The complacency it generates is in itself a bar to advance.

The fundamental reason why that advance was so long delayed was that in a feudal economy, Islamic or Christian, there was no way in which rational science could be used to any practical advantage. Astrology was esteemed enough by princes to keep astronomy going, and alchemy may have improved chemical technique, but owed little to reason, its theories being almost pure magic. As long as science was called on mainly to provide examples for theologians there was no reason to demand more than a formal analogy to experience. The searching test of practical use need never be applied. Science through the Middle Ages was accordingly largely confined to book learning and disputation. The intellectual advances that were to come later owed little to the schoolmen except the stimulus provided by the desire to prove them wrong. They were to come rather from the combination of the rediscovery of the best of classical thought with the new experimental methods inspired by a new practical interest in the world of Nature and art.

Much more significant for the future than medieval thought was the impressive total of technical development in manufacture and transport, and the legacy of difficult practical problems requiring the application of intelligence for their solution. The question raised at the outset as to what determined the time and place of the birth of modern science (pp. 55 f.) can be partly answered in terms of these considerations. Of the heirs to the first great burst of Hellenistic natural science, only western Europe was in a shape to make any forward move. By the fifteenth century the Islamic world had collapsed economically and had been ruined by internecine war and invasion. For all the later successes of the Turks and Mongols it had lost its intellectual drive. Its religion had ceased to be liberal and shrank to a narrow orthodoxy. India had become a battleground between waves of Islamic invaders and a Hinduism frozen in a caste structure that provided stability at the expense of any possibility of advance.

China preserved its old culture, but with a State system that prevented it, and would prevent it for another 400 years, from taking the necessary step of linking technique and book learning.

Culture in Europe at the end of the Middle Ages was hardly materially or even intellectually at a higher level than in the great empires of Asia. That it held greater promise could only be apparent from its relative lack of fixity and uniformity in social and economic forms (p. 1221). Great as was the weight of tradition, it was being everywhere challenged by consequences of the conflicts between the various interests of town and country, Church and State. Nor was the authority of pope and emperor, themselves most often at cross purposes, sufficient to impose any rigid limit on change. The feudal system itself that had given its essential character to the Middle Ages was evidently breaking up by the end of the fourteenth century. But this was not an evidence of social decay, for economically and technically there was in many places indubitable evidence of advance. If an old society was dying a new one was taking its place, one able to make far greater use of the advantages of the natural resources of Europe and the labour of its peoples than had the lords and prelates of the Middle Ages.

Table 3

Science and Feudalism: Salvaging the Hellenic Legacy (Chapters 5 and 6)

This table covers the 950 years from 500 to 1450. Throughout, the content of scientific thought – it would be hardly accurate to call it advance – is essentially Hellenic and indeed marks a direct continuation of that covered in Table 2. In contrast, the areas in which science was studied were far more widely scattered and the centres of interest in science varied with the time. Alexandria, Syria, Persia, Central Asia, India, China, were all active in the first part of the period; Spain, Italy, France, England, and the Low Countries in the latter part. It can be seen that, apart from a small burst of activity under Justinian, the three other significant but not major bursts occurred in Islamic Asia in the ninth century, in Islamic Spain in the eleventh, and in France in the thirteenth century. It is difficult to assign precise dates to the technical developments that were to be decisive in the next phase, such as the compass and gunpowder. All that can usually be indicated is the approximate date of introduction into Europe.

		Technical Developments	Political and Social Events
	— 500		
		St Sophia built	*Justinian* closes the Academy of Athens
		Silk introduced into Europe from China	Persian university of Jundishapur
	600		
			T'ang dynasty in China
		Block printing in China	*Mohammed*'s mission, spread of Islam to Persia, Africa, and Spain
	700		
		Use of wheeled plough and three-field system in Northern Europe	
			Abbasid caliphs
			Baghdad founded
	800		*Charlemagne*
		Vikings improve sailing-ships	*Haroun-al-Raschid*
		Introduction of horse-collar, shoes, and stirrups into Europe from China	Europe invaded by Northmen and Magyars
	900		
		Widespread use of water-mills	Break-up of caliphate
	1000	Windmills in Persia	Reform of Church
			Struggle of Pope and Emperor
		Use of lenses	
			Rise of Italian cities
		Alcohol	Invasion of Seljuk Turks
	— 1100	Paper in Spain	First Crusade
		Stained glass	Communes in Flanders
		Windmills in France	
		Mariner's Compass	*Saladin* retakes Jerusalem
	1200		
		Gunpowder introduced	*Frederick II* Emperor
		Villard de Honnecourt mechanical contrivances and clocks	
			Mongols sack Baghdad
	1300		
		Use of spectacles	Schism in papacy
		Cannons used in war	Hundred Years' War
		Sternpost ships	Black Death
		Oil painting	Peasant revolts
	1400		
			End of schism
		Printing	Hussite revolt
	— 1450		

The Transition to Feudalism Chapter 5

The Middle Ages Chapter 6

Dionysius mystical theology

Aryabhata ⎰Indian astronomers
Varahamihira ⎱and mathematicians

Philoponos anti-Aristotelian doctrine
   of impetus

Development of decimal numeration, the zero

Brahmagupta algebra and trigonometry

Severus Sebockt introduction of Hindu numerals into Syria

Translations from Greek into Syriac
Geber legendary founder of Islamic chemistry

Translations from Sanskrit, Syriac, and Greek into Arabic

House of Learning
Al-Kindi first Arab philosopher   *Al-Khwarizmi* algebra
Erigena first philosopher   *Bede* first historian of Christianity
Rise of Sufism, neoplatonist alchemical mysticism

Alfraganus founder of Islamic astronomy
Rhazes medicine and chemistry   *Al-Masudi* geometry
Abul Wafa trigonometry

Avicenna medicine and physics   *Al-Biruni* description of India
Alhazen founder of optics
Arzachel Toletan tables, elliptic orbits
Al-Ghazzali return to mysticism   *Omar Khayyam* mathematics

Peter Abelard University of Paris, beginning of scholasticism
Averroes Aristotelian Islamic system
Maimonides Aristotelian Jewish system

Translations from Arabic into Latin.   *Leonard of Pisa* introduction of Arabic numbers

Robert Grosseteste science in support of faith
Roger Bacon, Peter the Pilgrim experiment and science for use
St Albert, St Thomas Aquinas Aristotelian Christian system

Al-Tusi Ilkhanic Table   *Raymond Lull* Sufic mysticism and alchemy

Duns Scotus, William of Occam nominalism
Decay of scholasticism
Buridan, Oresme development of impetus doctrine

Ibn Khaldun science of history   *Ulugh Beg* Samarkand observatory
Nicholas of Cusa speculation on the earth's movement

Map 3

Medieval Europe

This map illustrates the distribution of towns and centres of learning in Medieval Christendom discussed in Chapter 6. It brings out the concentration on the central spine of Europe (p. 286 f.) and the two major trade routes of the Rhone and the Rhine running on each side of the Alpine barrier. Four areas are specially indicated as centres of economic revival: the North Sea area; the two Italian areas of Lombardy and Tuscany; the West and Mediterranean area of Provence and Languedoc which might be stretched to include Barcelona and the Balearic Islands. The embryonic industrial areas also marked are the copper and silver mines of Saxony, the tin mines of Cornwall, the iron mines of Styria, and the coalfields of Newcastle and Belgium.

Notes

(For explanation see page 23)

PAGE 41. *The distinction between the two modes goes very far back, if we follow the line of thought suggested by Professor Haldane and experiments on animal communication (p. 946). The mode of art appears as the most primitive, pre-human and almost pre-social. The magical, action-provoking use of language is used by birds. They do so by indicating their own internal state of emotion or preparation for action, often produced by events in the external world, but not in any sense describing them. This is, according to Haldane, the criterion of true human language in which a scientific element of description is superimposed on a magical call to action. The two modes shade off very gradually into each other. 'Let's go and fish' is still the magical-artistic mode. 'There were fish in that pool last year', which may follow it, is in the indicative-scientific mode. All these considerations are very general. There is no greater sense in giving a precise definition of art than there is of science. Both have had, in the course of history, an autonomous evolution. A novel is more akin to a scientific thesis than it is to a primitive ritual dance.

PAGE 52. *This also seems to be the opinion of that noted historian of science Thomas S. Kuhn. [1,25] In his *The Structure of Scientific Revolutions* he stresses what he has called the 'paradigm', a body of more or less self-consistent opinions, such as those of Aristotle or Galileo, each of which holds the field for varying periods and has then to be broken up and replaced by another. Though in my view he has largely concentrated on the ideological content of science and correspondingly less on the technological factors, this dialectical, though by no means admittedly Marxist, view of the history of science coincides very largely with my own, and is supported by a mass of detailed historical evidence.

PAGE 72. *The earliest stages of the development of tools and speech, though the most difficult to trace, are clearly the most important. I have already alluded (note p. 41) to Haldane's views of the pre-human origins of speech. He also thinks that some palaeolithic techniques may have been instinctive, like the making of birds' nests. As long as development is relatively slow this is not impossible by the kind of hereditary transmission of experience noticed in birds (p. 946). Once a critical stage is past, it seems to me that this mechanism, which is an excessively slow one, must have been taken over by social transmission of techniques which might be considered as the true origin of humanity.

PAGE 76. *Professor Haldane doubts the human origin of ritual. What we now know as such may be simply a verbalization of pre-human, or at least pre-linguistic action preparing and provoking dances. If true language came late, much of its social and economic function may have been originally taken by ritual.

PAGE 81. *Important aspects of primitive science may be overlooked simply because they are expressed in terms we no longer use. Descriptions and rules of actions may be expressed in terms of myth, but they are perfectly understandable in the framework of the particular culture. Thus, as de Santillana[2,3] has shown, Polynesian navigation achieved very accurate courses, though they were described in terms of star myths. In the same way, much mathematics may have been expressed in the patterns of sacred dances in the apparently abstract Chirunga drawings of the Australian Aborigines. We have been left one striking example of the numerical abilities of men of the early Stone Age in a bone which has been engraved with scratches indicating knowledge of the multiplication table and even of the existence of prime numbers. Many of the descriptions authors have given of the limited mathematical abilities of primitive man have exhibited not so much the ignorance of primitive man as our ignorance of him.

PAGE 86. *The importance of the shaman or medicine man for science has yet to be fully worked out. Certainly we owe to him the preservation, if not the discovery, of most useful drugs known before the twentieth century. The other magic arts, particularly of foreseeing the future, laid the foundation of the pseudo sciences of astrology and through them of the objective sciences of today (p. 334).

PAGE 93. *A small piece of supporting evidence for this is provided in a burial in one of the oldest human settlements in Persia at Siyalk, where two sheep jaws and a stone axe are the only grave goods.[2,32]

PAGE 93. †The possibility of agriculture turns on the fact that unlike meat grain is stored for a season once suitable vessels can be made. It may sprout in store, then die and ferment, but the effect of eating it in this state would also be appreciated, as it is on the Upper Nile today. Beer seems to be an even earlier human product than wine.

PAGE 96. *The excavations that have been carried out largely in the last decade and even in the last few years have pushed back our earlier conceptions of the date of the origin of neolithic culture. When the deeper levels are explored villages have been found that go back to dates a thousand or even two thousand years before those previously claimed. It would appear that agricultural civilization existed in various forms in Syria, Mesopotamia and Anatolia as far back as the eighth millennium B.C., not long after the end of the last ice age. What have been found are definitely the remains of villages with well-built houses depending on agriculture. Some appear to be more than villages; the remains of the earliest Jericho, for instance, include unmistakable fortifications. The most recently excavated of these also seem to be the oldest, those of Katal Küyük, near Konya in Turkey, excavated by Dr Melaart for instance. This is a village apparently of the pre-pottery neolithic but in most respects it is very advanced. The houses have plastered walls with frescoes in colour, mainly of animals, showing affinities to palaeolithic animal painting. There is also evidence of religious rituals and definite shrines. It is clear that whole new chapters have to be added to pre-history, but so far what has been found does not invalidate previous ideas as to the revolutionary effects of agriculture, making possible human concentrations in large numbers.

PAGE 105. *The control of the distribution of the grain stored in the temples is probably the earliest form of economic power. Later, other goods, particularly precious stones and metals, also came to be kept there, so that for many thousand years they effectively acted as banks. Where the king was also chief priest, as in ancient Egypt and China, the same function was fulfilled by the royal granaries. How important these could be in times of famines is well brought out in the story of Joseph.

PAGE 109. *This concentration of power in the hands of the few is essentially the social basis of civilization. Though it is made possible by technical developments, particularly those of food production, it in turn makes further technical developments possible and indeed necessary, as for example the fine handicrafts which were used for the benefit of the higher classes in the society, namely, fine pottery, fine weaving, etc. These arts have happily survived in most cases the breakdown of the unequal societies which gave rise to them. When a dynasty falls, craftsmen are scattered and end up in the villages passing on some of their skills. As the old Irish saying has it 'The castles are falling but the dunghills are rising.' This essentially social character of civilization is shown by the multiplicity of civilizations or part civilizations which have grown up pretty well independently in different parts of the world, notably in equatorial Africa and in America, as in Mexico and Peru. The specific characters or styles of the arts in these different places are clearly related to the previous tribal arts. Where they differ from them is in their concentration and the scale on which they are undertaken.[2.90]

PAGE 115. *An engraved seal from the third millennium in India shows such a bird in conjunction with a picture of a boat.

PAGE 115. †The plough seems to have had a double origin. It can be derived not only from the hoe but also from the digging stick dragged point forward through the ground. In either case the ground was just scratched or rooted up as if by a pig, hence many early names for plough and pig are identical. The deep-cutting, furrow-making plough was evolved from it much later.[2.19]

PAGE 116. *The existence of the wheeled vehicle would seem to imply roads, but these must have come much later. The first stage must have come in moderately level country with few water courses, and roads penetrated only later, if at all, into really hilly country. Paths for human beings or pack animals furnished the usual means of travel. In South China to this day, apart from the railways, most transport is by boat or human carriage; officials travel by palanquin.

PAGE 119. *The units of weight furnish the most imperishable record of early measurement, especially in the ratios they bear to each other, such as sixty minas to a shekel, the basis of Mesopotamian sexagesimal notation, also used in astronomy. The persistence of weight units, as are used today in Europe, which are basically of Roman or even earlier origin, is also evidence for the continuity of measurement throughout all the history of civilization.

PAGE 121. *The idea of the right angle certainly existed before building, however, and probably before weaving. Among the paintings on the walls of the caves at Lascaux are to be found so-called 'blazons', rectangular figures divided in somewhat irregular chess-board pattern with the squares alternately coloured. The most likely origin for this would seem to be plaiting, which we know was actually carried out in the Old Stone Age.

PAGE 123. *The invention of the week for religious purposes is shown to have had astronomical origin. Neugebauer[2.62] has traced its origin to the artificial quartering of the nearest approximation to a lunar month, ensuring automatically that it would get out of step with the real month of 28·9 days, and replacing a still older lunar calendar divided into three sets of nine days, associated with the moon goddess, the three-faced Hecate. The creation myth that fills the first part of the first chapter of Genesis is a priestly attempt to justify this astronomical construction, but it was the institution of the regular seventh-day sabbath which gave it really popular support.

PAGE 124. *Professor G. Thomson[2.84] has recently developed forcibly the arguments first put forward by Duhem that the origins of astronomy should be traced back to social totemic organization. This applies especially to the division of the sky into quarters corresponding to the splitting of the clan, each associated with appropriate totem animals and colours. By analogy this also explains the four elements, which, particularly in China, are associated with the quarters.[2.60a] The fifth central and royal element, the yellow earth, is a characteristic addition. It would seem as if the ideology or theory of science marks an adaptation to what is to us the inanimate world of concepts derived from society, just as the practice of science is achieved by the extension of the technical methods of early man. Sometimes, however, technique itself generates theory. In this particular case the wheel is transferred to the sky and identified with its motion as a whole, even in detail with the chariot of the sun, a great object of devotion in the Bronze Age. The mystique of the wheel with its motion of return is made into an image of human life and a promise of resurrection.

PAGE 153. *Professor G. Thomson[2.84] uses the criterion of commodity production drawn from Engels in *The Origins of the Family, Private Property and the State* to define civilization and thus by implication to refuse the term 'civilized' to cities before the Iron Age. He goes further to claim that this definition 'is superior to the traditional one, current among the bourgeois archaeologists, that civilization is the "culture of cities" '. To him writing and class division are as much general characteristics of civilization as urban development. In my opinion, as I think is evident from the context, where civilization is contrasted to communal clan society, Engels did not at all intend to limit civilization to its last and admittedly its most fully developed phase in the Greek city. It is true the 'culture of cities' definition is purely descriptive. What goes on in a city can be understood only in terms of productive relations and the essential division of labour, and separation of classes seems to be inseparable from the existence of the city in the first place. The physical form of the city and its social pattern grew up together, and, in my opinion, the resulting whole deserves the qualification of civilization which applies to Ur as much as Athens, though they mark two widely separated stages in its development.

PAGE 162. *It seems clear from recent excavations and discoveries that the continuity between the Greeks, the Cretans, and older Bronze Age civilizations was much closer than was formerly believed. Many detailed correspondences have been found between the culture of Crete and pre-Hittite Asia Minor. The most revolutionary discovery is that of Ventris, who showed that the late Cretans and Mycenaeans of 1500 B.C. wrote in Greek in a linear script derived from an earlier Cretan script.[2.87] The fact that there is little cultural break shows that these early Achaeans of Homer were to a large extent Cretized, with elaborate State and economic organization. The real break seems to have been with the second wave of Dorian (Spartan) Greeks in the tenth century, when the Cretan script seems to disappear to be replaced by the Phoenician. Even if over most of mainland Greece this indicates a dark age, there may have been no serious cultural break in the islands and the Anatolian shore, where in fact Greek science first appeared; Professor G. Thomson makes a very strong case for deriving Ionian science direct from Mesopotamian legends. Here it may well be that the essential condition which made the Greek renaissance possible was the impact on an old urban tradition of a much more primitive tribal society rapidly evolving towards city life.

PAGE 169. *This is one of the major questions which occupies historians of science in our day. As it stands it is intrinsically unanswerable. We cannot play the record of history over again with variations. Interesting new views, however, are to be found in the studies of Thomas S. Kuhn[1.25] and Giorgio de Santillana.[2.3]

PAGE 177. * The peculiar properties of the Platonic solids have exercised their influence through the whole of history. They were to prove an illusive guide, but a guide nevertheless, to Kepler in determining the system of the planets. They appeared again in modern science in such diverse fields as the structure of the viruses and the theory of liquids.[2.93]

PAGE 181. *The School of Pythagoras represents a first blending of number manipulation, essentially of Babylonian origin, with geometric form, which had already become a Greek speciality, though probably drawn from Egyptian sources. The introduction of number distinguishes geometry from mere drawing and makes it possible to introduce demonstrative logic, which is the essential Hellenic contribution to science. I strongly suspect, though I am not scholar enough to prove, that this decisive step was due to the transference to mathematics of the argued-out lawsuit which characterized the mercantile city states rather than the royal judgement. The notion of proof from postulates and even the argument by *reductio ad absurdum* all seem to stem from the courts. Q.E.D., *quod erat demonstrandum*, is the decisive plea which wins the case in the court of reason. Even though they were abused outside the realm of mathematics, the notions of proof and generality provided a guard against loose thinking which the science of ancient India and China never acquired.

PAGE 186. *The construction of a multi-spherical astronomy, as an extension of the simple wheeling of the heavens, was an achievement so familiar to the ancients and so forgotten by the moderns, that it is difficult to see what a great and decisive advance it was. Whereas to the precise Babylonian astronomers the movement of the skies was a matter of purely mathematical recurrences, facts to be reduced to numbers and formulae without any attempt at justification, the Greek view was frankly visual and almost tactile. They were attempting to reproduce a model in space of the celestial machine. They could hardly have thought of this if they had not been inspired by the existence of actual machines. We know now that clockwork was not new to the Greeks, and indeed in the pseudo-Aristotelian treatise *De Mundo* we find the author using a mechanical analogy to explain the action of God as a prime mover:

> For he needs no contrivance or the service of others, as our earthly rulers, owing to their feebleness, need many hands to do their work; but it is most characteristic of the divine to be able to accomplish diverse kinds of work with ease and by simple movement, even as past masters of a craft by one turn of a machine accomplish many different operations.

Thus he avoided the need for a detailed providence and opened a controversy that continues into our own time.

PAGE 187. *It is a sad commentary on the slowness of the progress of medical science and the obscurity of epilepsy that the causes of this particular disease are still almost wholly unknown, and nothing but palliative treatment can be given for it. Nevertheless, the study of epilepsy has thrown much light on the normal working of the brain.

PAGE 188. *This selection of colours is in fact far older than the doctrine of humours. They are the easiest colours to produce artificially and are the basic palette of stone age painting. They occur in Egyptian and later in Indian mythology, with some modification in Chinese as well. They are the four Varna which marked the distinction of the castes[2.46] (p. 194).

PAGE 207. *I have been accused, and rightly so, of having been unfair to Aristotle. My attitude would indeed be unpardonable if I had been writing a history of Greek philosophy or even of Greek science. However, for the purpose of this book, it is not so much what Aristotle actually thought or even wrote that is important, but rather what for so many centuries he was believed to have thought. The subtle and thought-provoking parts of Aristotle were not appreciated, the flat and vulgar parts remained. More charitably than I have inferred in the text, I would say that Aristotle's object was to give an accurate biological, that is natural, account of the world as a running concern. Even his logic, as Haldane points out, arises out of the difficulties of biological classification. I would not like to leave the impression that Aristotle's influence was consciously stultifying. Indeed, to the Arabs and even to schoolmen as late as Oresme in the fourteenth century it was a coherent vision of another world, an inspiration to ordered thinking. If St Thomas Aquinas entirely missed the spirit of inquiry and research in Aristotle it was not because he was incapable of it, but because he was a good Christian and would not follow that infidel Averroës.

PAGE 212. *I have been criticized for underestimating the contribution of Hellenistic free workers and their techniques to the advancement of science. It was certainly not my intention to do so. My major trouble here was lack of evidence, which can come only as a result of detailed studies which are either too scattered or have yet to be made. There is little material evidence for the achievements of the Hellenistic scientists and inventors. Some pumps and water clocks survive, as well as a pneumatic organ, but the chance find, recently restored by Dr Price, of a fragment from a sunken ship, shows that the Hellenistic technicians were able to produce most elaborate clockwork with variable connected gears, apparently not for the purpose of telling the time but for demonstrating the different movements of the planets.[3,51] From what I know already I can only conjecture that the great burst of technical improvements came in the first place from the fusion of techniques with those from Syria, Persia, and above all Egypt. The quick market for money-making devices – temple illusions, water-works, and clocks – must have had a stimulating effect on inventiveness similar to that of palace entertainments in the Renaissance, and the scientists for once were not uninterested. Not only did they learn the principles of hydrostatics and pneumotion but also produced improved and new devices themselves.

PAGE 216. *The influence of the old Babylonian mathematical and astronomical tradition – the two can hardly be separated – on Hellenistic science was probably greater than we allow for. Astronomically the continuity was acknowledged. Hipparchus quotes observations of Kidinnu. But the Greek geometric tradition was already set by the time of Alexander and the real fusion with the algebraic approach was achieved by the Arabs and not completed until the time of Descartes. It is evident from a few hints that have come down to us that intellectual activity continued in Babylon for at least a thousand years from the time of Nebuchadnezzar, under Persian, Greek, and, later, Persian domination up to Islamic times. Nor was this a mere preservation of tradition or continuance of observations, but real progress was made, such as the discovery of the precession of the ecliptic by Naburiannu in the fourth century B.C.

PAGE 222. *Professor Rosenfeld doubts this explanation for the failure of capitalist production to develop in classical times. He thinks that the cause was rather the lack of primary accumulation. I cannot agree with him. The loot of Asia must have provided as much treasure in the third century B.C. as the loot of the Indies, West and East, did in the sixteenth century. To my mind the reason is rather that practically none of the conditions which made capitalism possible in the seventeenth century existed in the Hellenistic world. The small manufacturers who made capitalism, if they had existed in Alexandria or Antioch, would have had little scope under a despotic government where money could be made so much more easily by government contracts or trade.

PAGE 230. *Professor E. A. Thompson draws attention to the statement of the anonymous fourth-century author of the *De rebus bellicis*[2,81] attributing the decay, visible in his time, to the seizure of the temple treasures by Constantine and their distribution not to the Church but to the Army. The money soon found its way to contractors and financiers and produced inflation and social unrest. These effects, which proved to be lasting, were an unexpected consequence of the triumph of Christianity.

PAGE 234. *This was achieved by the use of the heavy, eight-ox, celtic or teutonic *plough*, a modification of the original araraire or symmetrical plough (*n*. p. 115) achieved by removing one of the earth-scattering *ears* behind the share, turning the other into a *mould board* and adding on the plough beam a *coulter* to cut the sod and a plough foot to prevent the share cutting too deep. These were early replaced by a pair of wheels, hence the French term *charrue*, but the essential feature was the action in cutting a sod and turning it over in the *furrow*. It was difficult to turn, and this led to furlong-length strip cultivation instead of the cross-ploughed square fields suited to the light plough. This enormously important invention seems to be a genuine barbarian one; an example has been found in Denmark of the third century B.C.

PAGE 252. *I am very conscious of the bias of this work on the immediate ancestry of modern science towards the Classical, Islamic, and European–Medieval cultures. For completeness there should be an equivalent treatment of India and China. My excuse, however, is that we still lack the necessary analyses of the cultural and scientific histories of these countries and that it would be premature at this stage to say much more than I have about them. In China this work is actively being pursued by many scholars, and for the English reader there is Needham's monumental treatise.[3.8] For India, with its confused history and traditional reluctance to chronicle precisely, the task will be much harder and is only in fact beginning. In this connexion I have drawn on Professor D. D. Kosambi's history of Indian culture, which to my mind, at any rate, is the first account that begins to make sense.[2.46] The two great civilizations of India and China have remarkable similarities, as well as characteristic differences. Starting in the second millennium in a small area of the country, the Punjab and the middle Yellow River respectively, the Aryan and the early Han peoples gradually spread their culture by assimilation as much as by conquest over the whole subcontinent. The means was bureaucratic Confucian state organization in China and the Brahmin-dominated village system in India. However, India was much more exposed to foreign conquests and cultural influences than China, and as a result must be considered later as part of the Iranian Hellenistic Islamic culture group. China, with its parallel cultures in Japan, Korea, and Annam, does in fact form another cultural world exchanging ideas and devices but never till our own time forming part of the Hellenistic world.

PAGE 254. *Hinduism stands apart from other world religions by its absence, or rather by its multiplicity, of creeds. It is so inclusive in belief that it can contain complete materialistic atheism as well as primitive animism. The worshipper is left to his choice of gods and rituals. On the other hand, it is most exclusive as far as the worshipper is concerned. It is only for Hindus, and inside the fold it is indefinitely divided by *caste* and family. Owing to the unfortunate history of India in the last eight hundred years, it has been forced to remain essentially a religion of tradition and of acceptance of an increasingly unjust and unworkable social system. Only in this century, with liberation, comes the possibility of realizing the positive values of Hinduism and casting off the intricate net of traditions and dogmas as well as the crippling class divisions.

PAGE 255. *This interpretation has been enormously strengthened by the recent discoveries of the Dead Sea Scrolls,[3.21] the library of just such a community that flourished from 160 B.C. to A.D. 70. The rules of this community are specifically communistic. The worst curses are reserved for anyone who, like Ananias in the Acts of the Apostles, conceals his private wealth and fails to contribute it to the common stock. There are also many references to a persecuted Master of Righteousness. Whether this is the same as the historic Christ, or even if there is one, is still undetermined. Nor is this question today of the supreme importance that it has had over the centuries. What is indisputable now is that there were over the period in Judea a number of more or less spiritual, more or less political, reformers or resistance leaders. Round one or the other of these, or round a composite of all of them, could crystallize the hopes and loyalties of oppressed people brought up in the tradition of a Messiah and a dying and resurrected God.

Another illuminating revelation of the scrolls is the extent of Persian religious influence, with its contrast of powers of good and evil and its prospects of paradise. As the full story unfolds it will probably be seen that the Persian identification of religion with social justice has a continuous history inspiring Manichees, Mazdakists, Quarmatians, down to the Bahais of our own time, and has profoundly influenced the development of Christianity.[2.32]

PAGE 264. *The history of Indian science in the post-Gupta period is at present a gap in our knowledge that we may hope will not long remain unfilled. It seems to have been very largely a peninsular Indian development; the plains were too exposed to the type of barbarian invasions that affected Europe in the same period and destroyed its largely Graeco-Persian culture, culminating in the pillage of Mahmud of Gazni in the eleventh century. In the Deccan, however, and farther south there seems to have been a great meeting-place of cultures: Graeco-Roman, from the north and by the sea; Arab and Persian, also by sea; and Chinese carried by overland Buddhist pilgrims and oversea traders. Given these stimuli, it is not surprising that the native Indian genius in conditions of prosperity and cultural advance was able to produce a synthesis of knowledge which could in turn inspire early Islamic science. The chief scientific contributions seem to have been in mathematics and chemistry. The mathematical ideas show evidence in their form of earlier Chinese models, even more perhaps in their errors, but Indians seem also to have developed Babylonian ideas.[2.62]

PAGE 270. *The Persian contribution to world science has been inevitably underrated through the loss of books and lack of evidence.[2.32] In world history Persia should really be considered, together with central Asia and north-west India, as the eastern wing of the expansion of the old bronze age cultures of Egypt and Babylonia, corresponding to the western wing of Greece and Rome. The wars of the Greeks and Persians were effectively a great mixing of cultures which continued after the conquests of Alexander. All through the period of the European Dark Ages the empire of Persia was a bastion of civilization against the south-eastern wave of barbarians, much as the Roman Empire was against the north-eastern. Under the Sassanian emperors and the Abbasid Caliphs science was patronized and expanded. Only the use of Arabic language disguises the largely Persian element in Islamic science in the east, as it does the Graeco-Roman element in the west. The special contribution of the Persians seems to have been: in technology, the use of arch and dome construction and hydraulic engineering; in science, astronomical observation and the fusion of Greek and Babylonian with possibly some Chinese influence to build up algebra and trigonometry.

PAGE 280. *For a full account of the Chinese contribution to chemistry we must await the publication of the fifth volume of Needham's opus, but what he has already shown[2.60] has confirmed ideas as to general scientific method that are of the utmost importance in the history of science. The whole Chinese approach is what Needham calls the analogical attitude to science. It is an analysis of the whole of nature, not in terms of genesis or cause and effect – the main Babylonian–Greek approach – but rather as groups or bundles of things with some common characters. The five Chinese elements, for instance, equate water with shell-fish, black, salt, north, and winter; wood with fishes, green, sour, east, and spring; fire with birds, red, bitter, south, and summer; metal with beasts, acrid, white, west, and autumn; and the central earth with man, yellow, and sweet. An even more elaborate parallelism goes with the sacred trigrams ☰, ☱, etc. This way of accounting for everything in Nature in terms originally primitive and totemic is common enough in Western thought as in Aristotle's elements and the concept of macrocosm and microcosm. Although strictly nonsense, it can have its value in indicating the ways of science. Possibly in no other way could the multifarious phenomena of chemistry be comprehended; at any rate, it was along these lines, and not along those of causal mechanism, that chemical evolution did in fact progress. The same may also be true of biology; indeed, Needham qualifies the Chinese approach as organicist.

PAGE 283. *Irrigation agriculture is notably unstable, even after a short period of years. This is not so much due to the choking of canals, which can easily be cleared out, but to leakage from the canals and the consequent waterlogging of soil and the appearance of salt. A few failures in harvests lead to depopulation and when the population falls below the level needed to maintain the irrigation system, this becomes disastrous and the land reverts to desert. The beginnings of such a process are already to be seen in the agricultural irrigation works set up by the British in what is now Pakistan.

PAGE 287. *The East and West wings should also have been included in my account of medieval Christendom. The importance of the Elbe country from Bohemia to Schleswig grew with the opening up of the Slav lands to western trade and the exploitation of Baltic fisheries and timber. In Spain and Portugal the Moors were driven out step by step, and the new Christian kingdoms began to gain importance, for wool, metals, and sea trade.

PAGE 287. †How far the European feudal system was unique and why are questions of the utmost importance to the development of culture. The answers would probably show why backward Europe rather than any other part of the civilized world was to engender capitalism and science. My own guess is that its uniqueness depended on the fusion of the remains of order of the Roman empire, transmitted through the Church, with the vigorous and productive culture of the Northern barbarians, Franks, Germans, and Norsemen, whose clan organization had only just been broken by the corrupting influence of Roman wealth. Personal loyalty – for a consideration – took the place of clan solidarity and systematic exploitation of resources under the influence of the Church proved an improvement on simple robbery. A late form of the transition is beautifully set out in the Sagas, which mark the simultaneous conversion and infeudation of the Vikings, probably the most independent people in the world. Unbroken clan society, as in Ireland and the Highlands, proved far more resistant, and indeed in the Middle Ages managed to reabsorb the feudal Normans into clan organization. Yet the new forms were by no means imposed from above; they were adopted quite consciously by ambitious leaders to strengthen their positions and grudgingly accepted as a better alternative to a free life of intermittent pillage and murder. In the older civilized parts of the Empire, such as Italy, feudalism never got such a grip, nor did it in Islamic dominions, but militarily and even economically in the end these proved the weaker sections.

PAGE 288. *This account, though true for the latter stages of feudalism, cannot explain its origin. The original military service was essential for protection against external enemies, whether Norsemen, Magyars, Saracens or, finally, Mongols. Essentially it is of a military-technical character; the armoured knight and his horse was an effective answer to the light and much more mobile steppe horseman-archer. With stirrups and lance-rest he could drive through any opposition, though as the fourteenth century showed he was defeated by foot bowmen and was ultimately made completely obsolete by cannon. He could not protect all the peasants, but he could retire with some of them into a castle too strong to be taken except by long, mechanical siege. The knight was in effect the tank of those times. In the later Middle Ages he had outlived his military usefulness and had become mainly parasitic.

PAGE 305. *I am still, despite numerous critics, impenitent in my general judgement on the limited and unprogressive character of medieval science. I feel I could prove my point far better if I had more space, but to do so would be out of scale. I might, for instance, have said more about the impetus school, and particularly of Oresme, whom Professor Rosenfeld chides me for neglecting. Though I concede his importance as the first translator of science and philosophy into French, as Chaucer was to be in English, and for his contribution to economics in his adaptation of Aristotle in his *De mutationibus monetarum*, his original scientific contribution seems somewhat slender, being largely limited to a use of graphs and a correct definition of acceleration as uniformly difform motion. I did not rank him of cardinal importance because he was not consciously an innovator like Roger Bacon, and because his work was not followed up. This is no criticism of Oresme himself but of his times. Nothing much could be expected of France in the middle of the Hundred Years' War and the social unrest of the Jacquerie. Oresme stands more as the last of the Medievals rather than as a precursor of the Renaissance.

PAGE 308. *The same objective of reaching a comprehensive and reasonable account of nature and society and of the relations between them was achieved about the same time by the last of the great Chinese neo-Confucian philosophers, Chu Hsi (1131–1200). It also seems to be a concentration and rationalization of centuries of thought and tradition, including the mystical naturalistic Taoism and the formal conventional old Confucianism. It differs from the Western synthesis in its insistence on harmony rather than hierarchy by the complete absence of the idea of a directing God. In this it reflects the difference between the bureaucracy of China and the sovereignty of Christendom. Everything seems to find its explanation in terms of two generalized universals, Chhi and Li, almost untranslatable, which Needham gives as Matter-Energy and Organization respectively.[2,60a] From them emanate the Yang and the Yin (p. 280), the five elements (*n.*, p. 280), and all the rest of the Chinese scientific world-picture. These ideas, profound and vague, were later to have some effect on European philosophy through Leibniz. Perhaps, just because they lacked the obstinate wrong-headedness of the Aristotelian–Thomist synthesis, they were never to prove a stimulus to destructive criticism.

PAGE 310. *The systematic development of vaults, arches and buttresses, which led to an almost completely glass-covered surface of a church, must have helped the development of geometry. But Gothic architecture was in fact a dead-end. Some of its ideas may have lasted into the Renaissance in the tradition of accurate drawings, but it was the more aesthetic objective of perspective that was to lead to the greatest advance in mathematics.

PAGE 312. *Certainly the exploits of the great Chinese admiral Chêng Ho of the early fifteenth century – who sailed the Indian Ocean in fleets of large ships carrying thousands of men, conquered Ceylon, and explored the African coast – were deliberately stopped by the authorities. Technically the Chinese could easily have anticipated by a hundred years the exploits of Columbus and the Conquistadores. But they lacked the motive: Chinese economy did not need foreign trade. The most remarkable import was a giraffe, of which the Emperor disapproved. The Chinese Emperor had another and more pressing reason, the Tartar tribes in the northwest were still a serious peril and it needed all the resources of the Empire to hold them in check.

PAGE 317. *The work of Price[3.51] on interpreting bronze fragments from a ship that sank in the second century B.C. shows a complicated clockwork device which seems to be the kind of clock Archimedes is alleged to have made. In conjunction with Needham,[3.41] he has also reconstructed an elaborate Chinese tower clock of the eleventh century with gears and a mechanical escapement. This, by no means the first of its kind, was designed to regulate the Imperial protocol in accordance with the motions of the skies. The authors suggest that early clocks were not intended vulgarly to tell the time but rather as demonstrations of astronomic principles to princes. What happened in the Middle Ages was first their use as town clocks, which made them common, and then their use in astronomy and navigation, which demanded greater accuracy.

Bibliography to Volume 1

PART 1

1. ANTHONY, H. D., *Science and its Background*, 4th ed., London, 1962
2. BERNAL, J. D., *The Freedom of Necessity*, London, 1949
3. BERNAL, J. D., *Science and Industry in the Nineteenth Century*, London, 1953
4. BERNAL, J. D., *The Social Function of Science*, London, 1939
5. BLACK, M., 'The Definition of Scientific Method', *Science and Civilization*, ed. R. C. Stauffer, Wisconsin, 1949
6. BUNGE, M. A., *Causality: the Place of the Causal Principle in Modern Science*, Cambridge, Mass., 1959
7. CALDER, R., *The Inheritors. The Story of Man and the World He Made*, London, 1961
8. CHILDE, V. G., *History*, London, 1947
9. CHILDE, V. G., *What Happened in History*, Penguin Books, 1942
10. CONANT, J. B., *On Understanding Science*, New Haven, 1947
11. COPE, Sir Z. (ed.), *Sidelights on the History of Medicine*, London, 1957
12. CROMBIE, A. C. (ed.), *Scientific Change*, London, 1963
13. CROWTHER, J. G., *The Social Relations of Science*, London, 1941
14. DAMPIER, Sir W. C., *Cambridge Readings in the History of Science*, Cambridge, 1924
15. DAMPIER, Sir W. C., *A History of Science and its Relations with Philosophy and Religion*, Cambridge, 1949
16. DAMPIER, Sir W. C., *Shorter History of Science*, Cambridge, 1944
17. DERRY, T. K., and WILLIAMS T. I., *A Short History of Technology from the Earliest Times to A. D. 1900*, Oxford, 1960
18. DINGLE, H., *The Scientific Adventure*, London, 1952
19. DUGAS, R., *A History of Mechanics*, trans. J. R. Maddox, London, 1957
20. ECO, U., and ZORZOL, G. B., *A Pictorial History of Inventions: from Plough to Polaris*, London, 1962
21. FORBES, R. J., *Man the Maker: a History of Technology and Engineering*, London, 1959
22. HULL, L. W. H., *History and Philosophy of Science*, London, 1959
23. JORDAN, P., *Science and the Course of History*, London, 1956
24. KLEMM, F., *A History of Western Technology*, trans. D. W. Singer, London, 1959
25. KUHN, T. S., *The Structure of Scientific Revolutions*, Chicago, 1962
26. LEICESTER, H. M., *The Historical Background of Chemistry*, London, 1956
27. LEVY, H., *The Universe of Science*, London, 1933
28. LILLEY, S. 'Social Aspects of the History of Science', *Archives Internationales d'Histoire des Sciences*, vol. 28, 1949
29. MASON, S. F., *A History of the Sciences*, London, 1953

30. PARTINGTON, J. R., *A History of Chemistry*, vol. II, London, 1961
31. POLANYI, M., *Science, Faith and Society*, London, 1946
32. PRICE, D. J. DE S., *Little Science, Big Science*, New York, 1963
33. ROUSSEAU, P., *Histoire de la science*, Paris, 1945
34. SARTON, G., *The History of Science and the New Humanism*, New York, 1931
35. SARTON, G., *On the History of Science*, London, 1962
36. SCOTT, J. G., *A History of Mathematics from Antiquity to the Beginning of the Nineteenth Century*, London, 1958
37. TATON, R., *Histoire générale des sciences*, 3 vols., Paris, 1957–62
38. TATON, R., *Reason and Chance in Scientific Discovery*, trans. A. J. Pomerans, Paris, 1957
39. SINGER, C., *A Short History of Science to the Nineteenth Century*, Oxford, 1941
40. SINGER, C., et al., *A History of Technology*, 5 vols., Oxford, 1954–8
41. TAYLOR, F. S., *Science Past and Present*, London, 1945
42. TAYLOR, F. S., *A Short History of Science*, London, 1949
43. TOY, S., *A History of Fortification from 3000 B.C. to A.D. 1700*, London, 1955
44. TOYNBEE, A. J., *A Study of History*, 6 vols., Oxford, 1939
45. TURNER, D. M., *The Book of Scientific Discovery*, 3rd ed., London, 1960
46. UNDERWOOD, E. A. (ed.), *Science, Medicine and History*, 2 vols., Oxford, 1953
47. UNESCO, *The History of Mankind: Cultural and Scientific Development*, 7 vols., 1963–
48. WHEWELL, W., *History of the Inductive Sciences*, 3 vols., London, 1857
49. WHITEHEAD, A. N., *Science and the Modern World*, Cambridge, 1925
50. WIGHTMAN, W. P. D., *The Growth of Scientific Ideas*, Edinburgh, 1950

51. BELL, E. T., *The Development of Mathematics*, New York, 1945
52. CLARK, G. N., *Science and Social Welfare in the Age of Newton*, 2nd ed., Oxford, 1949
53. DINGLE, H., 'Science and Professor Bernal', *Science Progress*, no. 146, London, 1949
54. DOIG, P., *A Concise History of Astronomy*, London, 1950
55. EINSTEIN, A., *The World as I See It*, London, 1935
56. FARADAY, M., *Faraday's Diary*, ed. T. Martin, 8 vols., London, 1932–6
57. FARRINGTON, B., 'Karl Marx, Scholar and Revolutionary', *Modern Quarterly*, vol. 7, 1952, p. 83
58. FORBES, R. J., *Man the Maker*, 2nd ed., London, 1958
59. GREGORY, J. C., *A Short History of Atomism*, London, 1931
60. HAWTON, H., *The Feast of Unreason*, London, 1952
61. LILLEY, S., *Men, Machines and History*, London, 1948
62. MUMFORD, L., *The Culture of Cities*, London, 1940
63. MUMFORD, L., *Technics and Civilization*, London, 1947
64. NORDENSKIÖLD, E., *The History of Biology*, New York, 1928
65. PARTINGTON, J. R., *A Short History of Chemistry*, London, 1948
66. RAYLEIGH, LORD, *The Life of Sir J. J. Thomson*, Cambridge, 1942
67. ROLL, E., *A History of Economic Thought*, London, 1938
68. RUSSELL, B. A. W., *A History of Western Philosophy*, London, 1946
69. SIGERIST, H. E., *A History of Medicine*, vol. 1, New York, 1951
70. SINGER, C., *A History of Biology*, London, 1950
71. SINGER, C., *A Short History of Medicine*, Oxford, 1928
72. SMITH, D. E., *A History of Mathematics*, 2 vols., Boston, 1923, 1925

73. STRUIK, D. J., *A Concise History of Mathematics*, 2 vols., New York, 1948
74. THORNDIKE, L., *A History of Magic and Experimental Science*, 8 vols., New York, 1923–58
75. USHER, A. P., *A History of Mechanical Inventions*, 2nd ed., Cambridge, Mass., 1954

76. *Ambix*, London, 1937–
77. *Annals of Science*, London, 1936–
78. *Archives Internationales d'Histoire des Sciences*, Paris, 1947–
79. *Arts and Sciences in China*, London, 1963–
80. *AScW Journal*, London, 1955–
81. *British Journal for the History of Science*, Ravensmead, 1962–
82. *British Journal for the Philosophy of Science*, Edinburgh, 1950–
83. *Bulletin of the Atomic Scientists*, Chicago, 1946–
84. *Bulletin of the British Society for the History of Science*, London, 1949–
85. *Centaurus*, Copenhagen, 1950–
86. *History of Science*, Cambridge, 1962–
87. *Isis*, Brussels, 1913–
88. *Journal of Biophysical and Biochemical Cytology*, New York, 1955–
89. *Journal of the History of Ideas*, New York, 1940–
90. *Labour Monthly*, London, 1920–
91. *Marxist Quarterly*, London, 1954–
92. *Modern Quarterly*, London, 1938–53
93. *Notes and Records of the Royal Society*, London, 1938–
94. *Operational Research Quarterly*, London, 1950–
95. *Osiris*, Bruges, 1936–
96. *Revue d'Histoire des Sciences et de Leur Application*, Paris, 1947–
97. *Science and Mankind*, London, 1949–
98. *Science and Society*, New York, 1936–
99. *Science for Peace Bulletin*, London, 1951–
100. *The New Scientist*, London, 1956–
101. *The Scientific Worker*, London, 1920–54
102. *Scientific American*, 1845–
103. *Scientific World*, London, 1957–
104. *Transactions of the Newcomen Society*, London, 1922–

PART 2

1. BRUNET, P., and MIELI, A., *Histoire des Sciences: Antiquité*, Paris, 1935
2. CHILDE, V. G., *Man Makes Himself*, London, 1939
3. DE SANTILLANA, G., *The Origins of Scientific Thought: from Anaximander to Proclus, 600 B.C. to 300 A.D.*, London, 1962
4. FARRINGTON, B., *Science and Politics in the Ancient World*, London, 1939
5. FARRINGTON, B., *Science in Antiquity*, London, 1936
6. REY, A., *La Science dans l'antiquité*, 4 vols., Paris, 1930–46
6a. SARTON, G., *Ancient Science to Epicurus*, Cambridge, Mass., 1952

7.   SARTON, G., *Introduction to the History of Science*, vol. I, Baltimore, 1927
8.   SINGER, C., *From Magic to Science*, London, 1928

9.   ADCOCK, Sir F. E., *The Greek and Macedonian Art of War*, London, 1962
10. ANDREWS, E., *A History of Scientific English*, New York, 1947
11. ARISTOTLE, *Aristotle's Physics*, ed. W. D. Ross, Oxford, 1936
12. BAILEY, C., *The Greek Atomists and Epicurus*, Oxford, 1928
13. BREASTED, J. H., *The Edwin Smith Surgical Papyrus*, Chicago, 1930
14. BUDGE, E. A. W., *Egyptian Hieratic Papyri in the British Museum*, 2nd series, London, 1923
15. CICERO, *Tusculan Disputations*, ed. and trans. J. E. King, London, 1927
16. CLAGETT, M., *Greek Science in Antiquity*, London, 1957
17. COLLINGWOOD, R. G., *The Idea of Nature*, Oxford, 1945
18. CORNFORD, F. M., *The Unwritten Philosophy and other essays*, Cambridge, 1950
19. CURWEN, E. C., *Plough and Pasture*, London, 1946
20. DANGE, S. A., *India from Primitive Communism to Slavery*, Bombay, 1949
21. DIAMOND, A. S., *The Evolution of Law and Order*, London, 1951
22. DUHEM, P., *Le Système du monde*, 5 vols., Paris, 1913–17
23. ENGELS, F., *Dialectics of Nature*, trans. and ed. C. Dutt, London, 1940
24. FARRINGTON, B., *Greek Science*, vol. I, Penguin Books, 1944
25. FARRINGTON, B., *Greek Science*, vol. II, Penguin Books, 1949
26. FINLEY, M. I., *The World of Odysseus*, London, 1956
27. FORBES, R. J., *Metallurgy in Antiquity*, Leiden, 1950
28. FORBES, R. J., *A Short History of the Art of Distillation*, Leiden, 1948
29. FRANKFORT, H., *The Birth of Civilization in the Near East*, London, 1951
30. FRANKFORT, H., et al., *Before Philosophy*, Penguin Books, 1949
31. GALEN, *De sanitate tuenda*, trans. R. M. Green, Oxford, 1952
32. GHIRSHMAN, R., *Iran*, Penguin Books, 1954
33. GLANVILLE, S. R. K. (ed.), *The Legacy of Egypt*, Oxford, 1942
34. GRANET, M., *Chinese Civilization*, London, 1930
35. HAMMURABI, *The Oldest Code of Laws in the World*, Edinburgh, 1905
36. HAUDRICOURT, A. G., and HEDIN, L., *L'Homme et les plantes cultivées*, Paris, 1935
37. HEATH, Sir T. L., *Aristarchus of Samos*, Oxford, 1913
38. HEATH, Sir T. L., *Greek Astronomy*, London, 1932
39. HEATH, Sir T. L., *A History of Greek Mathematics*, Oxford, 1921
40. HEIBERG, J. L., *Mathematics and Physical Science in Classical Antiquity*, trans. D. C. Macgregor, Oxford, 1952
41. HEIDEL, W. A., *The Heroic Age of Science*, Washington, 1933
42. HYAMS, E., *Soil and Civilization*, London, 1952
43. JAFFE, B., *Crucibles: the Story of Chemistry from Ancient Alchemy to Nuclear Fission*, London, 1959
44. KENYON, K. M., 'Early Jericho', *Antiquity*, vol. 26, 1952
45. KING, L. W., *The History of Sumer and Akkad*, London, 1916
46. KOSAMBI, D. D., *An Introduction to the Study of Indian History*, Bombay, 1956
47. KOYRÉ, A., *La Révolution astronomique*, Paris, 1961
48. KUHN, T. S., *The Copernican Revolution: Planetary Astronomy in the Development of Western Thought*, London, 1957
49. LANDSTROM, B., *The Ship: A Survey of the History of the Ship from the Primitive Raft to the Nuclear-Powered Submarine*, London, 1962

50. LASLETT, P. (ed.), *Philosophy, Politics and Society*, Oxford, 1956
51. LEFEBVRE DES NOETTES, R., *De la Marine antique à la marine moderne*, Paris, 1935
52. LEFEBVRE DES NOETTES, R., *L'Attelage*, Paris, 1931
53. LEROI-GOURHAN, A., *L'Homme et la matière*, Paris, 1943
54. LEROI-GOURHAN, A., *Milieu et techniques*, Paris, 1945
55. LINDNER, K., *La Chasse préhistorique*, Paris, 1950
56. LIVINGSTONE, Sir R. W. (ed.), *The Legacy of Greece*, Oxford, 1942
57. LLOYD, S., *Early Anatolia*, Penguin Books, 1956
58. LOVEJOY, A. O., *The Great Chain of Being*, Oxford, 1936
59. MASON, O. T., *The Origins of Invention*, London, 1895
60. MINS, H. F., 'Marx's Doctoral Dissertation', *Science and Society*, vol. 12, 1948
60a. NEEDHAM, J., 'L'Unité de la Science', *Archives Internationales d'Histoire des Sciences*, no. 7, 1949
61. NEUBURGER, A., *The Technical Arts and Sciences of the Ancients*, London, 1930
62. NEUGEBAUER, O., *The Exact Sciences in Antiquity*, 2nd ed., Providence, R.I., 1957
63. OAKLEY, K. P., *Man the Tool-maker*, 5th ed., London, 1961
64. PALLOTTINO, M., *The Etruscans*, Penguin Books, 1955
65. PARTINGTON, J. R., *A History of Greek Fire and Gunpowder*, Cambridge, 1960
66. PARTINGTON, J. R., *Origins and Development of Applied Chemistry*, London, 1935
67. PIGGOTT, S., *Prehistoric India*, Penguin Books, 1950
68. PLATO, *Dialogues*, trans. B. Jowett, 3rd ed., 5 vols., Oxford, 1951
69. PLINY, the Elder, *Naturalis historia*, English translation in 10 vols., London, 1951–
70. PLUTARCH, 'The Life of Marcellus', *Plutarch's Lives*, trans. B. Perrin, vol. 5, London, 1914–26
71. POPPER, K. R., *The Open Society and its Enemies*, London, 1945
72. POPPER, K. R., 'The Nature of Philosophical Problems and their Roots in Science', *British Journal for the Philosophy of Science*, vol. 3, 1952
73. RAGLAN, LORD, *The Hero*, London, 1949
74. ROBERTSON, A., *The Bible and its Background*, 2 vols., London, 1949
75. SAMBURSKY, S., *The Physics of the Stoics*, London, 1959
76. SANDARS, N. K. (trans.), *The Epic of Gilgamesh*, Penguin Books, 1960
77. SHAPIRO, H. L. (ed.), *Man, Culture and Society*, London, 1956
78. SINGER, C., *Greek Biology and Greek Medicine*, Oxford, 1922
79. SPEISER, E. A., 'The Beginnings of Civilization in Mesopotamia', Supplement to *Journal of the American Oriental Society*, no. 4, 1939
80. THEOPHRASTUS, *Theophrastus's History of Stones*, trans. J. Hill, London, 1746
81. THOMPSON, E. A., *A Roman Reformer and Inventor*, Oxford, 1952
82. THOMSON, G., *Aeschylus and Athens*, London, 1946
83. THOMSON, G., *Studies in Ancient Greek Society*, London, 1949
84. THOMSON, G., *Studies in Ancient Greek Society, vol. 2: The First Philosophers*, London, 1955
85. THOMSON, G., 'From Religion to Philosophy', *Journal of Hellenic Studies*, vol. 73, 1953
86. VAILLANT, G. C., *The Aztecs of Mexico*, Penguin Books, 1950
87. VENTRIS, M., and CHADWICK, J., *Documents in Mycenaean Greek*, Cambridge, 1956
88. VON HAGEN, V. W., *Highway of the Sun*, London, 1956
89. WASON, M. O., *Class Struggles in Ancient Greece*, London, 1947

90. WELTFISH, G., *The Origins of Art*, New York, 1953
91. WHEELER, Sir M., *Rome Beyond the Imperial Frontier*, Penguin Books, 1955
92. WHITE, L. A., *The Evolution of Culture: the Development of Civilization to the Fall of Rome*, New York, 1959
93. WHYTE, L. L., *Essay on Atomism: from Democritus to 1960*, London, 1961
94. WITTFOGEL, K. A., *Wirtschaft und Gesellschaft Chinas*, Leipzig, 1931

PART 3

1. ARBERRY, A. J. (ed.), *The Legacy of Persia*, Oxford, 1953
2. BUTTERFIELD, H., *The Origins of Modern Science*, 2nd ed., London, 1957
3. BURCKHARDT, J., *The Civilization of the Renaissance in Italy*, London, 1944
4. BURNS, C. D., *The First Europe*, London, 1947
5. CROMBIE, A. C., *From Augustine to Galileo*, London, 1952
6. HEER, F., *The Medieval World*, trans. J. Sondheimer, London, 1962
7. MIELI, A., *La Science arabe*, Leiden, 1939
8. NEEDHAM, J., *Science and Civilization in China*, 4 vols., Cambridge, 1954–62
9. PIRENNE, H., *Economic and Social History of Medieval Europe*, London, 1949
10. SAMBURSKY, S., *The Physical World of Late Antiquity*, London, 1962
11. SARTON, G., *Introduction to the History of Science*, vols. 2 and 3, Baltimore, 1931, 1947
12. WHITE, L., *Medieval Technology and Social Change*, Oxford, 1962

13. ARNOLD, Sir T. W., and GUILLAUME, A. (eds.), *The Legacy of Islam*, Oxford, 1931
14. ARTZ, F. B., *The Mind of the Middle Ages*, New York, 1953
15. ASIN PALACIOS, M., *Islam and the Divine Comedy*, London, 1926
16. BACON, R., *Essays on Roger Bacon*, ed. A. G. Little, Oxford, 1914
17. BACON, R., *Opus Majus*, trans. R. B. Burke, 2 vols., Philadelphia, 1928
18. BASHAM, A. L., *The Wonder That was India*, London, 1954
19. BAYNES, N. H., and MOSS, ST L. B. (eds.), *Byzantium*, Oxford, 1948
20. BOËTHIUS, *The Consolation of Philosophy*, trans. H. R. James, London, 1897
21. BURROWS, M., *The Dead Sea Scrolls*, London, 1956
22. CARTER, T. R., *The Invention of Printing in China and its Spread Westward*, New York, 1931
23. CHAUCER, G., 'A Treatise on the Astrolabe', *Early English Text Society*, *Extra Series 16*, London, 1872
24. CLAGETT, M., *The Science of Mechanics in the Middle Ages*, Madison, 1959
25. CLOW, A. and N., *The Chemical Revolution*, London, 1952
26. CROMBIE, A. C., *Robert Grosseteste*, Oxford, 1953
27. EASTON, S. C., *Roger Bacon*, London, 1952
28. GARREAU, A., *Saint Albert le Grand*, Paris, 1932
29. GAUTHIER, L., *Ibn Rochd (Averroës)*, Paris, 1948
30. GIBBS, M., *Feudal Order*, London, 1949
31. GILFILLAN, S. C., *Inventing the Ship*, Chicago, 1935
32. GRECOV, B. D., *The Culture of Kiev Rus*, Moscow, 1947

33. GUNTHER, R. W. T., *Early Science in Oxford*, 14 vols., Oxford, 1923–45
34. HASKINS, C. H., *Studies in the History of Medieval Science*, Cambridge, Mass., 1927
35. HITTI, P. K., *A History of the Arabs*, 4th ed., London, 1949
36. HOLMYARD, E. J., *Alchemy*, Penguin Books, 1957
37. IBN KHALDUN, *Selections from the Prolegomena of Ibn Khaldun of Tunis* (1332–1406), trans. and arr. C. Issawi, London, 1950
38. IYYUBH (JOB) OF EDESSA, *Book of Treasures*, trans. A. Mingana, Cambridge, 1935
39. LINKLATER, E., *The Ultimate Viking*, London, 1955
40. NEEDHAM, J., *Chinese Science*, London, 1950
41. NEEDHAM, J., et al., 'Chinese Astronomical Clockwork', *Nature*, vol. 177, 1956
42. NEEDHAM, J., et al., *Heavenly Clockwork: the Great Astronomical Clocks of Medieval China*, Cambridge, 1960
43. O'LEARY, DE L., *How Greek Science Passed to the Arabs*, London, 1948
44. ORESME, N., *Le Livre de Ethiques d'Aristotle*, Critical Introduction and Notes by A. D. Menut, New York, 1940
45. PEERS, E. A., *Fool of Love: the Life of Ramón Lull*, London, 1946
46. PEREGRINUS, P., 'Epistola de Magnete', *Proc. Brit. Acad.*, vol. 2, 1905–6
47. PIRENNE, H., *Histoire économique de l'Occident médiéval*, Bruges, 1951
48. PIRENNE, H., *Medieval Cities*, Princeton, 1925
49. PIRENNE, H., *Mohammed and Charlemagne*, trans. B. Miall, London, 1940
50. POWER, E. E., *The Wool Trade in English Medieval History*, London, 1941
51. PRICE, D. J., 'Clockwork before the Clock', *Horological Journal*, vols. 97 and 98, 1955 and 1956
52. PRICE, D. J., 'The Equatorie of the Planetis', *Bull. Brit. Soc. Hist. Sci.*, vol. 1, 1953
53. RASHDALL, H., *The Universities of Europe in the Middle Ages*, 3 vols., Oxford, 1936
54. READ, J., *Prelude to Chemistry*, London, 1936
55. RENAN, J. E., *Averroës et l'Averroïsme*, Paris, 1866
56. ROBERTSON, A., *The Origins of Christianity*, London, 1953
57. ROBERTSON, J. D., *The Evolution of Clockwork*, London, 1931
58. SINGER, C., *The Earliest Chemical Industry*, London, 1948
59. STENTON D. M., *English Society in the Early Middle Ages*, Penguin Books, 1951
60. TAYLOR, F. S., *The Alchemists, Founders of Modern Chemistry*, New York, 1949
61. THOMAS AQUINAS, ST, *Summa Theologica*, London, 1913–42
62. WAITE, A. E., *Three Famous Alchemists*, London, 1939
63. WALBANK, F. W., *The Decline of the Roman Empire in the West*, London, 1946
64. WINTER, H. J. J., *Eastern Science*, London, 1952

Note on the Illustrations

The choice of illustrations for Professor Bernal's *Science in History* has been based on the simple principle of providing additional illumination of the text. Since the author has taken so wide a canvas on which to display his analysis, the range of illustrations has been made as broad as possible. However, science has not always been illustrated at every stage in its history, and from some periods, of the few illustrations which may have existed, little or no evidence has survived to the present day, in consequence certain problems had to be solved if gaps were to be avoided. For example, virtually no original material remains of Greek science, and the scientific texts that we have are copies or translations made in later centuries. In such cases, later sources have been used if, as often happens, they make the point; Greek ideas continued for so long in western Europe that it is often still valid to use material from printed books.

In this book, where both science and the interplay of social conditions are discussed, the pictures could not always be chosen as direct illustrations of the text, but in every case it is hoped that the full captions will enable the reader to see why a picture has been chosen and appreciate its relevance, whether as allusion or analogy, by comparison or even as a comment. No attempt has been made to illustrate Professor Bernal's introductions to the various sections of this book, since this would have caused too great a mixture of subjects and historical periods. By confining illustration to the main body of the text, some degree of chronological order has been possible.

The choice of each picture has depended on a number of factors: its relevance to the text, the quality of the illustration itself, its power to provide additional visual or factual information and, of course, its aesthetic appeal. Here and there diagrams have been used, but in every case they are of historical significance. In volume 1, except for the need to cast the net wide for material about Greek science, the illustrations are comparatively straightforward. Volume 2 has almost illustrated itself. Volume 3, dealing primarily with modern scientific research, is again straightforward, but volume 4 has presented some problems, in that its theme – the social sciences in history – is so wide, and that some of the concepts cannot be illustrated directly. The solution adopted has been to try, in one way or another, to complement the spirit of the text. Sources of illustrations have been given wherever possible, in a separate acknowledgements section on p. 363.

My thanks are due to Mr Francis Aprahamian for his helpful advice, and especially to my wife, whose assistance and extensive library of illustrations has proved invaluable.

COLIN A. RONAN
Cowlinge, Suffolk
June 1968

Acknowledgements for Illustrations

For permission to use illustrations in this volume, acknowledgement is made to the following: the Trustees of the British Museum for numbers 1a, 1b, 4, 18, 20, 51, 85; Dreyers Forlag, 7 (*d, e*); John R. Freeman & Co. Ltd, 64; the Griffith Institute, Ashmolean Museum, 28; the Louvre, 29; Lund Humphreys, London 87a and 87b; the Mansell Collection, 3, 8, 12, 16, 17, 25, 32, 33, 34, 36, 40b, 41, 43, 52, 56, 62, 63, 70, 75, 79, 80, 83; the Ministry of Public Buildings and Works, Scotland, 67; the Trustees of the National Gallery, 72 and 99; the National Museum, Denmark, 2; Dr Joseph Needham, 88; the Palestine Archaeological Museum, 38; Paul Popper Ltd, 6, 14, 15, 21, 26, 35, 44, 73; R.B.O. for number 13; the Ronan Picture Library, 9, 10, 22, 24, 27, 30, 31, 37, 40a, 42, 45, 47, 48, 55, 61, 65, 66, 74, 78, 81, 82, 84, 89, 92, 93, 95, 96, 97, 98; the Ronan Picture Library and E. P. Goldschmidt & Co. Ltd, 46, 60, 71, 76, 86, 90, 94; the Ronan Picture Library and the Royal Astronomical Society, 5, 50, 53, 54, 57, 58, 59, 68, 100; the Trustees of the Royal Scottish Museum, 77; the Science Museum, London, 23 and 91; Bernard G. Silberstein, 39; the Smithsonian Institute, 7 (*a, b, c*); the Trustees of the Victoria and Albert Museum, 69; the Wellcome Trustees for number 49.

The publishers would like to thank Mrs Sheila Waters for drawing maps 1, 2 and 3.

Name Index

Bold figures indicate main reference

Subject Index